TRANSCEIVER SYSTEM DESIGN FOR DIGITAL COMMUNICATIONS

by

Scott R. Bullock. P.E.

Noble Publishing
Atlanta

Library of Congress Cataloging-in-Publication Data

Bullock, Scott, 1950-
   Transceiver system design for digital communications / by Scott R. Bullock
     p.   cm.
   Includes bibliographical references and index.
   ISBN 1-884932-46-0
   1. Radio--Transmitter-receivers--Design and construction.
2. Spread spectrum communications. 3. Digital communications.
I. Title
TK6561.B835   1995
621.384'131--dc20   95-33738
                                                                                          CIP

To order contact:
                Noble Publishing
                2245 Dillard Street
                Tucker, Georgia 30084
                USA
                TEL (770)908-2320
                FAX (770)939-0157

Discounts are available when ordered in bulk quantities.

Technical Editor: Gary Breed
Cover Designer and Copy Editor: Crawford Patterson

NOBLE PUBLISHING

© 1995 by Noble Publishing
All rights reserved. No part of this book may be reproduced in any form or by any means without written permission of the publisher. Contact the Permissions Department at the address above.

Printed and bound in the United States of America
10 9 8 7 6 5 4 3 2 1

International Standard Book Number 1-884932-46-0
Library of Congress Catalog Card Number 95-33738

To Debi, my loving wife, and

To Crystal, Cindy, Brian, Andy, and Jenny

Contents

Contents

Preface

- Transceiver Design 1
    - 1.1 Frequency of Operation 2
    - 1.2 The Link Budget 3
    - 1.3 Power in dB 5
    - 1.4 Transmitter 9
    - 1.5 Channel 16
    - 1.6 Receiver 21
    - 1.7 Summary 43
    - 1.8 References 45
- The Transmitter 49
    - 2.1 Antenna 49
    - 2.2 Transmit/Receive Control 51
    - 2.3 Upconversion 52
    - 2.4 LO and Elimination of Sideband 54
    - 2.5 Power Amplifier 59
    - 2.6 VSWR 60
    - 2.7 Spread Spectrum Transmitter 61
    - 2.8 PN Code Generator 86

2.9	Summary	88
2.10	References	89

The Receiver ... 91

3.1	Superheterodyne Receiver	92
3.2	Antenna	93
3.3	Transmit/Receive Control	94
3.4	Limiters	94
3.5	Image Reject Filter	95
3.6	Dynamic Range/Minimum Discernable Signal	96
3.7	Types of Dynamic Range	100
3.8	Second and Third Order Intermodulation Products	103
3.9	Calculating Two Tone Frequency Dynamic Range	106
3.10	System Dynamic Range	109
3.11	Tangential Sensitivity	114
3.12	Low Noise Amplifier	116
3.13	Downconversion	119
3.14	Splitting Signals into Multiple Bands for Processing	120
3.15	Phase Noise	120
3.16	Mixers	122
3.17	Bandwidth Constraints	129
3.18	Filter Constraints	130
3.19	Pre-Aliasing Filter	131
3.20	A/D Converter	133
3.21	Digital Signal Processing	136
3.22	Summary	137
3.23	References	137

AGC Design And PLL Comparison 139
   4.1   AGC Design 140
   4.2   AGC Amplifier Curve 142
   4.3   Linearizer 144
   4.4   Detector 144
   4.5   Loop Filter 149
   4.6   Threshold Level 151
   4.7   Integrator 151
   4.8   Control Theory Analysis 152
   4.9   Modulation Frequency Distortion . 156
   4.10 Comparison of the PLL and AGC
          Using Feedback Analysis Techniques . 159
   4.11 Basic PLL 159
   4.12 Comparisons of the PLL and AGC ... 161
   4.13 Detector 163
   4.14 Loop Filter 164
   4.15 Loop Gain Constant 164
   4.16 Integrator 165
   4.17 Conversion Gain Constant 165
   4.18 Control Theory Analysis 165
   4.19 Summary 171
   4.20 References 173

Demodulation 177
   5.1   Pulsed Matched Filter 178
   5.2   Matched Filter Correlator 179
   5.3   Pulse Position Modulation 183
   5.4   Code Division Encoding 185
   5.5   Coherent Demodulation 186
   5.6   Despreading Correlator 187
   5.7   Carrier Recovery 190
   5.8   Symbol Synchronizer 198

5.9	The Eye Pattern	199
5.10	Digital Processor	202
5.12	Phase Shift Detection	204
5.13	Summary	208
5.14	References	209

Basic Probability and Pulse Theory 213
- 6.1 Simple Approach to Understanding Probability 213
- 6.2 The Gaussian Process 220
- 6.3 Quantization Error 223
- 6.4 Probability of Error 228
- 6.5 Probability of Detection and False Alarms 231
- 6.6 Pulsed System Probabilities using Binomial Distribution Function 235
- 6.7 Error Detection and Correction 236
- 6.8 Sampling Theorem and Aliasing 237
- 6.9 Theory of Pulse Systems 237
- 6.10 PN Code 240
- 6.11 Summary 243
- 6.12 References 244

Multipath 247
- 7.1 Basic Types of Multipath 248
- 7.2 Specular Reflection on a Smooth Surface 248
- 7.3 Specular Reflection on a Rough Surface 251
- 7.4 Diffuse Reflection 253
- 7.5 Curvature of the Earth 256
- 7.6 Pulse Systems (RADAR) 257

7.7	Vector Analysis Approach	259
7.8	Power Summation Approach	260
7.9	Alternative Approach	262
7.11	References	267

Improving the System Against Jammers 271
8.1	Burst Jammer	272
8.2	Adaptive Filter	279
8.3	Digital Filter Intuitive Analysis	281
8.4	Basic Adaptive Filter	282
8.5	LMS Algorithm	285
8.6	Digital/Analog ALE	288
8.7	Wideband ALE Jammer Suppressor Filter	296
8.8	Digital Circuitry	298
8.9	Simulation	299
8.10	Results	299
8.11	Gram-Schmidt Orthogonalizer (GSO)	305
8.12	Basic GSO	308
8.13	Adaptive GSO Implementation	311
8.14	Intercept Receiver Comparison	314
8.15	Summary	319
8.16	References	323

Global Navigation Satellite Systems 327
9.1	Satellite Transmissions	328
9.2	Data Signal Structure	329
9.3	GPS Receiver	331
9.4	Atmospheric Errors	331
9.5	Multipath Errors	332
9.6	Narrow Correlator	335
9.7	Selective Availability	337
9.8	Carrier Smoothed Code	338

9.9	Differential GPS	340
9.10	DGPS Time Synchronization	341
9.11	Relative GPS	341
9.12	Doppler	342
9.13	Carrier Phase Tracking	342
9.14	Double Difference ($\Delta\nabla$)	344
9.15	Wide Lane/Narrow Lane	346
9.16	Summary	346
9.17	References	347

Direction Finding and Interferometer Analysis		351
10.1	Interferometer Analysis	351
10.2	Direction Cosines	352
10.3	Basic Interferometer Equation	355
10.4	Three Dimensional Approach	359
10.5	Antenna Position Matrix	361
10.6	Coordinate Conversion due to Pitch and Roll	364
10.7	Using Direction Cosines	366
10.8	Alternate Method	371
10.9	Summary	372
10.10	References	373

Preface

This was written for those who want a good understanding of how to design a spread spectrum transceiver and a good intuitive and practical approach.

This text will provide a good resource for anyone involved in transceiver design for digital communications. This text also includes a basic understandings of spread spectrum in general, and many of the aspects of the actual design. This text is geared more towards basic direct sequence analysis but many of the principles can be applied to other forms of spread spectrum or digital communications.

The text starts off in Chapter 1 with the transceiver design using a link budget to analyze the tradeoffs and to track the power and noise levels in the system. This chapter includes the calculation of noise figure, the gains and losses of the link, and the required signal to noise level including the link margin to provide the specified probability of error.

The transceiver is separated into the transmitter and the

receiver designs. In Chapter 2 the transmitter is evaluated including front end design and sideband elimination discussion in the up conversion process. Several spread spectrum techniques using phase shift keying (PSK) are provided along with block diagrams and phasor diagrams to help analyze the different types of PSK systems that are used today. Variations of the PSK systems and other types of spread spectrum systems are also discussed, including frequency hopping, time hopping and chirped FM.

The receiver design is presented in Chapter 3. This chapter covers the design requirements of designing a receiver which includes dynamic range, both amplitude and frequency, two-tone dynamic range, different receiver architectures, phase noise and mixer spurious signal analysis. This chapter also covers filters and bandwidth constraints, pre-aliasing filters, and analog/digital (A/D) converters including piecewise linear A/Ds.

In Chapter 4, an extensive design and analysis of automatic gain control (AGC) and the elements of a good AGC design is provided including the amplifier curve, linearizer, detector, the loop filter, threshold level, and the integrator. Control theory is used to define the stability characteristics of the AGC in order to design the optimal AGC for the system. Chapter 4 also analyzes the phase lock loop (PLL), particularly for the lock condition and makes a comparison of the similarities between the AGC and the PLL.

Chapter 5 describes the demodulation process portion of the receiver which includes the different methods of correlating the incoming digital waveform. This includes the matched filter, the coherent sliding correlator, pulse position modulation and demodulation, code tracking loops including the early-late gate analysis, and the autocorrelation function. This chapter also contains different types of carrier recovery loops including the squaring loop, Costas loop, and modifications of the Costas loop. Also included in Chapter 5 is a discussion of the symbol synchronizer, the eye pattern, intersymbol interference (ISI), and phase-shift detection for intercept receivers.

In Chapter 6, a basic discussion on the principles of understanding digital communications is provided. This includes a good intuitive and analytical approach to understanding probability theory when used in designing and analyzing digital communications. Included in the discussion is the basic gaussian distribution and how to apply this to probability of error. Quantization error is also evaluated for the system and the probability of error for different types of spread spectrum systems are provided along with the curves and how to apply them in a design. Also discussed is the probability of false alarm and probability of detect, checksum for detecting errors, and sampling theory and aliasing. Chapter 6 also provides basic theory on pulsed systems which includes spectral plots of the different pulse types.

Multipath is the main topic of Chapter 7. This chapter discusses the basic types of multipath including specular reflection off of both smooth and rough surfaces and diffuse reflections off a glistening surface. This includes the Rayleigh criteria for determining if the reflections are specular or diffuse. The curvature of the earth is included for systems such as those used for satellite communications. In Chapter 7 the advantages of using leading edge tracking for radars is discussed to mitigate most of the multipath present. Several approaches to the analysis of multipath is provided including a vector analysis approach and a power summation approach.

In Chapter 8, methods are described that improve the system operation against jamming signals. The methods include burst clamps to minimize the effects of burst jammers, adaptive filtering to reject narrowband signals, and a Gram-Schmidt orthogonalizer which uses two antennas to suppress the jamming signal. An in depth analysis is provided on the adaptive filter method using the adaptive filter configured as an adaptive line enhancer using the least mean square (LMS) algorithm. This discussion describes an actual wideband system providing results of the filter. A discussion of different types of intercept receivers is also provided fo detection of signals in the spectrum.

Since Global Positioning Service (GPS) is a direct sequence spread spectrum data link, Chapter 9 provides a discussion of the GPS system. This includes data signal structure, receiver characteristics, narrow correlation, selective availability (SA), carrier smoothing of the code, kinematic carrier phase tracking (KCPT), the double difference, and wide lane verses narrow lane techniques.

Chapter 10 provides direction finding and interferometer analysis using direction cosines and coordinate conversion techniques to provide the correct solution. This chapter provides a discussion of the drawbacks to the standard interferometer equation, and details the necessary steps to design a three-dimensional interferometer solution.

Special thanks to Larry Huffman, senior scientist and system engineer, for thoroughly reviewing this text and enhancing the book with his experience and expertise.

Thanks to Don Shea, a recognized expert in the field of antennas and interferometers, for providing the technical expertise and consulting for Chapter 10.

Scott R. Bullock received his BSEE degree from Brigham Young University in 1979 and his MSEE degree from University of Utah in 1988. He is a licensed Professional Engineer and a member of IEEE and Eta Kappa Nu.

He currently holds the position of Senior Engineering Specialist for the Montek Division of E-Systems, a Raytheon Company, where he specializes in data link design and system analysis.

He has published several articles dealing with spread spectrum modulation types, multipath, AGCs, PLLs, and adaptive filters. He has been doing research and development work for most of his career. His experience and research activities are mainly in the area of digital transceivers for data link communications.

1

Transceiver Design

The Transceiver is a term that is used to describe a communication link that has a transmitter that transmits the signal through space on one end and a receiver to receive the signals on the other end as shown in Figure 1-1.

Proper transceiver design is critical in the cost and performance of a data link. In order to provide the optimal design for the transceiver, a link budget is used to allocate the gains and losses in the link.

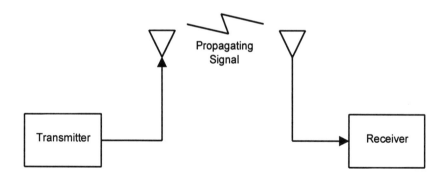

Figure 1-1 *The transceiver block diagram.*

1.1 Frequency of Operation

Generally the first thing to determine in a transceiver design is the RF frequency of operation.

This depends on several factors, some of which are specified below:

- a. RF Frequency Availability -- this is the frequency band that is available for use by a particular system and is specified by the Federal Communications Commission (FCC) which has the ultimate control over the frequency band allocation.

- b. Cost -- as the frequency increases, the components in the receiver tend to be more expensive. An exception to the rule is when there is a widely used frequency band, such as with the cellular radio band, where supply and demand drives down the cost of parts and where integrated circuits are designed for specific applications.

- c. Size and Range -- as a general rule, as the frequency decreases, the free-space attenuation decreases which results in an increase in range or a lower power output requirement which would affect cost. Also, the size of the antenna increases as the frequency decreases. This could affect siting constraints due to the size of the antenna and could also be a factor in the cost of the design.

d. Customer Specified -- often the frequency of operation is specified by the customer. This needs the approval of the FCC.

e. Band Congestion -- ideally the frequency band selected is an unused band or has very little use, especially with no high power users in the band. This also needs to be approved by the FCC.

Taking into consideration the above criteria, and other factors, the frequency of operation is selected. Once the frequency has been chosen, a link budget is performed to aid in the design of the transceiver.

1.2 The Link Budget

The link budget is a term used to determine the necessary parameters for a successful transmission of a signal from a transmitter to a receiver through space. The term 'link' refers to linking or connecting the transmitter to the receiver which is done by means of sending out radio frequency (RF) waves through space. The term 'budget' refers to the allocation of RF power, gains and losses, and noise throughout the entire system including the link between the transmitter and the receiver. The main items that are included in the budget are the required power output level from the transmitter power amplifier, the gains and losses throughout the system and link, and the signal-to-noise level required at the receiver to produce the desired bit error rate or the probability of detection and probability of false alarm. Therefore, when certain parameters are known or selected, the link budget allows the system designer to calculate unknown parameters.

Several of the link budget parameters are given or chosen during the process and the rest of the parameters are calculated. There are many variables and tradeoffs in the design of a transceiver, and each one needs to be evaluated for each system that is being design. For example, there are tradeoffs between the power output required from the power amplifier and the size of the antenna. The larger the antenna (producing more gain) the less power is required from the power amplifier. However, the cost and size of the antenna may be too great for the given application. On the other hand, the cost and size of the power amplifier increases as the power output increases and this may be the limiting factor. If the power output requirement is large enough, a solid state amplifier may not be adequate and therefore a traveling wave tube amplifier (TWTA) may be needed. The TWTA requires a special high voltage power supply which generally increases size and cost. Therefore, by making these kinds of trade studies, an optimum solution can be found for any of the systems that are being designed.

Before starting the link budget, all fixed or specified information concerning the transceiver needs to be examined to determine which parameters to calculate in

Transceiver Design

the link budget. The tradeoffs need to be evaluated before the link budget is performed and then re-evaluated to ensure that the right decisions have been made. The parameters for a link budget are listed and explained in this chapter.

1.3 Power in dB

The term dB is the abbreviation for decibels. The main reason that dB is used is to enable the engineer to calculate the resultant power level by simply adding or subtracting gains and losses instead of multiplying and dividing. For example:

Amplifier input = 150 µW = –8.2 dBm

Amplifier power gain = 13 = 11.14 dB

Power output = 150 µW × 13 = 1.95 mW = 2.9 dBm

Power output in dB = –8.2 dBm + 11.14 dB = 2.9 dBm = 1.95 mW

The term dB is used in the industry extensively and it has been wrongly used in many cases. To clear up some misunderstanding, the following is an explanation of different dB terms:

    a.   The term dB is a change in signal level, signal amplification or signal attenuation. It is the ratio of power output to power input. It can also be the difference from a given power level such as, "so many dB down from a reference." The equation is:

$$dB = 10\log\frac{P_o}{P_i} \qquad 1.1$$

where:

P_o = power out

P_i = power in

b. For voltage, the same equation is used substituting V^2/R for P:

$$dB = 10\log\frac{(V_o^2/R_o)}{(V_i^2/R_i)} \qquad 1.2$$

where:

V_o = voltage out

V_i = voltage in

R_o = output impedance

R_i = input impedance

If $R_o = R_i$ then:

$$dB = 10\log\frac{V_o^2}{V_i^2} = 10\log(\frac{V_o}{V_i})^2 = 20\log\frac{V_o}{V_i} \qquad 1.3$$

Many times dB, which is a ratio of powers, is mistakenly used in the place of dBw, which is actually a power level related to watts. Another misconception is that there is not a dB for voltage separate from a dB for power; a dB is a dB regardless of whether voltage or power is used.

Therefore, if the system has 6 dB of gain, it is the same for voltage or power. The only difference is that the ratio of voltage is increased by two and the power ratio is increased by four to give the overall system gain of 6 dB. For example, if an amplifier has 6 dB of gain, it has 4 times more power and 2 times more voltage (given that $R_i = R_o$). If $R_i \neq R_o$, then the power ratio should always be used since the voltage ratio only causes confusion.

The following are some definitions for several log terms that are referred to and used in the industry today:

1. dB is ratio of signal levels, which indicate the gain or loss in signal level.

2. dBm is a power level in milliwatts, 10logP where P = power (milliwatts).

3. dBw is a power level in watts, 10logP where P = power (watts).

4. dBc is how far down the signal strength is with respect to the carrier in dB.

5. dBi is the gain of an antenna with respect to the gain of an isotropic radiator.

There are several other dB terms, similar to the last two, that refer to how many dB away a particular signal is from a reference level. For example, if a signal is 20 times larger than a reference signal r, then the signal is 13 dBr, which means that the signal is 13 dB larger than the reference signal.

There is one more point to consider when applying the term dB in a link budget. When referring to attenuation or losses, the output power is less than the input power so the attenuation in dB is negative. Therefore, when calculating the link budget, the attenuation and losses are added.

Transceiver Design

1.4 Transmitter

The transmitter is the part of the transceiver that creates, modulates and transmits the signal through space to the receiver (see Figure 1-1). The transmitter is responsible for providing the necessary power required for transmitting the signal through the link to the receiver. This includes the power amplifier, the transmitter antenna, and the gains and losses associated with the process, such as cable losses, to provide the effective radiated power (ERP) out of the antenna (see Figure 1-2).

1.4.1 Power from Transmitter

The power from the transmitter is the amount of power output of the final stage of the power amplifier:

P_t = transmitter power

For ease of calculating the link budget, the power is in dBm or converted to dBm from milliwatts.

Power in milliwatts is converted to power in dBm by:

$$P_{dBm} = 10\log_{10} P_{milliwatts} \qquad 1.4$$

Therefore, one milliwatt is equal to 0 dBm. The unit dBm is used extensively in the industry and a good understanding of this term and other dB terms is important. The term dBm is actually a power level related to one milliwatt and is not a loss or gain as in the term dB.

1.4.2 Transmitter Component Losses

With most systems, there will be RF components (circulators, transmit/receive switch, etc.) that enable the transceiver to use the same antenna for both transmit and receive. Also, if there are antenna arrays, (multiple antennas), there will be components that interconnect the individual antenna elements. Since these elements have

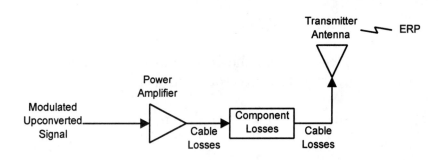

Figure 1-2 The transmitter block diagram.

a loss associated with them, they need to be taken into account when calculating the link budget. Often times this loss is included in the noise figure of the system.

The alternate method, which the author prefers, is to not include the losses before the low noise amplifier (LNA) in the receiver as part of the noise figure. Those losses merely reduce the signal with respect to the noise and do not contribute additional noise. The losses reduce the signal for the link budget directly. Therefore, the component losses are labeled and included in the analysis as shown below:

$$L_{tcomp} = (\text{switching, circulator, etc.})$$

Which ever method is used, the losses directly affect the link budget on a one-for-one basis. One dB loss equals one dB loss in link margin. Therefore, the losses between the final output power amplifier at the transmitter and the first amplifier of the receiver should be kept to a minimum. Each dB of loss in this path will either reduce the minimum discernable signal (MDS) by one dB or the power amplifier will have to transmit one dB more power.

1.4.3 Transmitter Line Losses from Power Amplifier to Antenna

Since most transmitters are located a distance from the antenna, the cable and/or waveguide connecting the transmitter to the antenna contains losses that need to be incorporated in the link budget. Therefore:

L_{tll} = Coaxial or waveguide line losses in dB

These transmitter line losses are included in the link budget as a direct attenuation, one dB loss equals one dB loss in link margin. Using larger diameter or better cables can reduce the loss which is a tradeoff with cost. Also, locating the transmitter power amplifier as close as possible to the antenna will reduce this loss.

1.4.4 Transmitter Antenna Gain

Most antennas experience gain since they tend to focus the energy in certain directions as opposed to an isotropic antenna which radiates in all directions. Even a simple dipole antenna experiences approximately 2.1 dBi of gain. A parabolic dish radiator is commonly used at high frequencies to achieve gain in the direction the antenna is pointing. The gain for a parabolic antenna is given by [1]:

$$G_t = 10\log[n(\pi(D)/\lambda)^2] \qquad 1.5$$

where:

n = efficiency factor < 1.
D = diameter of the parabolic dish.
λ = wavelength

Transceiver Design

Notice that the antenna gain increases both with increasing diameter and higher frequency (shorter wavelength). The gain of the antenna is a direct gain in the link budget. One dB gain equals one dB improvement in link margin. Therefore, the more gain the antenna can produce, the less power the power amplifier has to deliver. Here again, is another tradeoff that needs to be considered to ensure the best design and the lowest cost.

1.4.5 Transmitter Antenna Losses

There are several losses associated with the antenna. Some of the possible losses, which may or may not be present in each antenna are listed below:

a. L_{tr} = radome losses on transmitter antenna -- the radome is the covering over the antenna to protect the antenna from the outside elements that can cause losses.

b. L_{tpol} = polarization loss of antenna -- many antennas are polarized, horizontal, vertical, or circular. This defines the spatial position or orientation of the electric and the magnetic fields. A loss is inherent due to polarization, with the loss being greater for horizontally polarized antennas as compared to vertically polarized antennas.

c. L_{tcon} = conscan cross over loss -- this loss is present if the antenna is scanned in a circular search pattern.

d. L_{tfoc} = focusing loss or refractive loss -- this is a loss when the antenna receives signals at low elevation angles.

e. L_{tpoint} = mispointed loss -- this is caused by transmitting and receiving directional antennas that are not exactly lined up. Therefore, the gains of the antennas do not add up without a loss of signal power.

The total losses are included in the link budget:

$$L_{ta} = L_{tr} + L_{tpol} + L_{tcon} + L_{tfoc} + L_{tpoint} \qquad 1.6$$

These losses are also a direct attenuation, 1 dB loss equals 1 dB loss in link margin.

1.4.6 Transmitted Effective Isotropic Radiated Power

The power transmitted out of the antenna in the correct direction is related to the power coming from an isotropic radiator. Therefore, to analyze the output of the antenna, an Effective Isotropic Radiated Power (EIRP) is used:

$$\text{EIRP} = P_t + L_{tcomp} + L_{tll} + G_t + L_{ta} \qquad 1.7$$

Transceiver Design

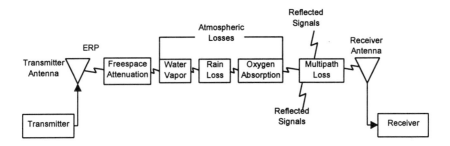

Figure 1-3 *The channel attenuation block diagram.*

This is also referred to as the Effective Radiated Power, (ERP), and combines the output power of the transmitter, line losses and the losses and the gain of the antenna.

1.5 Channel

The channel is the path of the RF signal that is transmitted from the transmitter antenna and received by the receiver antenna (see Figure 1-1). This is the signal in space that is attenuated by the channel medium. The main contributor to the channel loss is known as the freespace attenuation. The other factors including the propagation losses, multipath losses etc., are fairly small compared to freespace loss. The losses are depicted in Figure 1-3.

1.5.1 Freespace Attenuation

As a wave propagates through space, there is a loss associated with it. This loss is due to dispersion, the "spreading out" of the beam of radio energy as it propagates through space. This loss is consistent relative to wavelength, which means that it increases with frequency, as the wavelength becomes shorter. This is called free space loss or path loss and is related to both the frequency and the slant range, the distance between the transmitter and receiver antennas. The equation is given by:

$$A_{fs} = 20\log[\lambda/(4\pi R)] = 20\log[c/(4\pi Rf)] \qquad 1.8$$

where:

λ = wavelength.
R = slant range, same units as λ.

Transceiver Design

c = speed of light, 300×10^6 meters/sec, R is in meters.
f = frequency

At first glance, the free-space loss appears to decrease as the range increases, however, as the number inside the brackets gets smaller and is less than one, the resultant becomes a larger negative number. Therefore, the free-space loss increases as the both the range and the frequency increase. This loss is also a direct attenuation, 1 dB loss equals 1 dB loss in link margin.

1.5.2 Propagation Losses

There are other losses, depending on the conditions in the atmosphere, such as clouds, rain, humidity, etc. that need to be included in the link budget. The three main losses are:

   a.   Cloud Loss -- loss due to water vapor in clouds

   b.   Rain Loss -- due to rain, dependent on operating region and frequency.

   c.   Atmospheric Absorption (Oxygen Loss) -- loss due to the atmosphere.

These values are usually obtained from curves and vary from day to day and from region to region. Each application is dependent on the location and a nominal loss, generally not worst case, is used for the link analysis.

To get a feel for the types of attenuation that can be expected, some typical numbers are shown below:

>Rain Loss = –1 dB/mile at 14 GHz with .5"/hr. rainfall

>Rain Loss = –2 dB/mile at 14 GHz with 1"/hr. rainfall

>Atmospheric Losses = –.1 dB/nautical mile for a frequency range of approximately 10-20 GHz

>Atmospheric Losses = –.01 dB/nautical mile for a frequency range of approximately 1 to 10 GHz.

A typical example gives a propagation loss of approximately –9.6 dB for the following conditions:

>Frequency = 18 GHz
>Range = 10 nm
>Rainfall Rate = 12 mm/hr
>Rain temperature = 18 degrees C
>Cloud Density = .3 grams/m^3
>Cloud Height = 3-6 kilometers

To determine the actual loss for a particular system, there are several excellent sources that can be refered to. Crane's

Transceiver Design

[2] model for rain is a good tool to use for this type of analysis.

This loss is also a direct attenuation for the link budget, one dB loss equals one dB loss in link margin.

1.5.3 Multipath Losses

Whenever a signal is sent out in space, the signal can either travel on a direct path or it can take multiple indirect paths. The most direct path the signal can take is with the least amount of attenuation. It is similar to a pool table, where you can hit a ball by aiming directly at the ball or you can bank it off the table with the correct angle to hit the ball.

The problem with multipath is that the signal takes both paths and interferes with each other at the receiving end. The reflected path has a reflection coefficient that can change the phase and amplitude, and the path length gives the reflected signal a different phase. If the phase of the reflected path is different, say 180 degrees out of phase, from the direct path, and the amplitudes are the same, then the signal is cancelled out and the receiver sees very little to no signal.

Most of the time the reflected path is attenuated depending on the reflection coefficient and the type of multipath. It does not completely cancel out the signal but can alter the amplitude and phase of the direct signal. Therefore, multipath can affect coverage and accuracy where reliable amplitude or phase measurements are required. A further discussion on multipath is presented in Chapter 7. The losses are included in the link budget as follows:

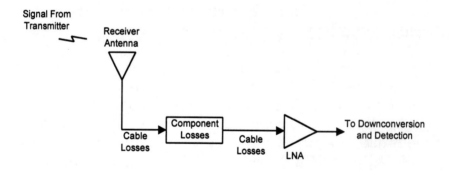

Figure 1-4 *The receiver block diagram.*

Transceiver Design

L_{multi} = Losses due to multipath cancellation of direct path signal in dB.

This loss is generally hard to quantize and is usually associated with a probability number, for example two sigma. The multipath is constantly changing and certain conditions can adversely affect the coverage and the phase measurement accuracy. This loss is also a direct attenuation, 1 dB loss equals 1 dB loss in link margin. Careful siting of the antennas in a given environment is the most affective way to reduce the effects of multipath. Also using blanking methods to ignore signals received after the desired signal for long multipath returns is often used to help mitigate multipath. Further detailed discussion on multipath will be addressed in Chapter 7.

1.6 Receiver

The receiver accepts the signal that was sent by the transmitter through the channel via the receiver antenna (see Figure 1-4). The losses from the antenna to the low noise amplifier (LNA) should be kept as small as possible. This includes cable losses and component losses that are between the antenna and the LNA (see Figure 1-4). The distance from the receiver antenna and the LNA should be kept as small as possible. Some systems actually include the LNA with the antenna system, separate from the rest of the receiver, to minimize this loss. The main job of the receiver is to receive the transmitted signal and detect the data that it carriers in the most effective way without further degrading the signal.

1.6.1 Receiver Antenna Losses

There are antenna losses for the receiver that are very similar to those for the transmitter. Some of the commonly occurring losses are listed below:

a. L_{rr} = radome losses on receiver antenna - the radome is the covering over the antenna to protect the antenna from the outside elements, and can cause losses.

b. L_{rpol} = polarization loss of receiver - many antennas are polarized, horizontal, vertical, or circular. This defines the spacial position or orientation of the electric and the magnetic fields. A loss is realized due to polarization.

c. L_{rcon} = conscan cross over loss - this loss is present if the antenna is scanned in a circular search pattern.

d. L_{rfoc} = focusing loss or refractive loss - this is a loss when the antenna receives signals at low elevation angles.

e. L_{rpoint} = mispointed loss - this is caused by transmitting and receiving directional antennas that are not exactly lined up. Therefore the gains of the antennas do not add up without a loss of signal power. Note that this loss may be combined into one number so that it is not included in both the receiver analysis and the transmitter analysis.

The total losses for the antenna can be calculated by adding all of the losses together, assuming that their values are in dB:

$$L_{ra} = L_{rr} + L_{rpol} + L_{rcon} + L_{rfoc} + L_{rpoint} \qquad 1.9$$

This total loss, as was the case in the transmitter section, is a direct attenuation of the signal, 1 dB loss equals 1 dB loss in link margin.

1.6.2 Receiver Antenna Gain

The gain of the receiver antenna in dB is calculated in the same way as the transmitter antenna gain:

$$G_r = 10\log[n(\pi(D)/\lambda)^2] \qquad 1.10$$

where:

n = efficiency factor < 1.
D = diameter of the parabolic dish.
λ = wavelength

The receiver antenna is not required to have the same antenna as the transmitter. The receiver could use an omni-directional antenna and receive the transmissions from a transmitter using a parabolic dish antenna, or the transmitter could use an omni-directional antenna and the receiver could use a parabolic dish. The gain of the antenna is a direct gain in the link budget, 1 dB gain equals 1 dB improvement in link margin.

1.6.3 Receiver Line Losses from Antenna to Low Noise Amplifier (LNA)

The cable that connects the antenna to the first amplifier which is designated as a Low Noise Amplifier (LNA), is included in the budget:

L_{rll} = Coaxial or waveguide line losses in dB.

The amplifier is referred to as a LNA since the noise figure of the system is determined mainly by the first amplifier in the receiver. This amplifier is designed to have a very low noise figure. The calculation and discussion of the noise figure of a system will be provided in more detail further in the chapter. The important thing is that the noise figure directly affects the link budget as a one for one dB loss. The cable loss between the antenna and the LNA, as was the case in the transmitter section, is a direct attenuation of the signal, 1 dB loss equals 1 dB loss in link margin. Therefore, as mentioned before, this cable length should be kept as short as possible, with the option of putting the LNA with the antenna assembly.

Transceiver Design

1.6.4 Receiver Component Losses

Any components between the antenna and LNA reduce the signal-to-noise of the system. For example, often times a limiter is placed in the line between the antenna and the LNA to protect the LNA from being damaged by high power signals. This can be looked at as an increase in noise figure when discussing the receiver only. Also, it could be evaluated by calculating the attenuation that the components contribute before the LNA. The noise stays the same since the temperature through the devices is constant. The noise is kTB noise, where;

    k = Boltzman's Constant
    T = Nominal Temperature (290 degrees)
    B = Bandwidth

The LNA amplifies the noise above the kTB noise. In this process, noise is added which is the noise figure of the LNA. Following the LNA, gain/loss components are applied to both the signal and the noise unless the loss is so great that the noise starts approaching the kTB noise again. Also, if the bandwidth becomes larger than the previous kTB bandwidth initially established, the noise can increase and degrade the noise figure. The kTB noise is a constant noise floor unless the temperature or bandwidth is changed.

Therefore in the link budget, the receiver component losses before the LNA are applied to the signal thus reducing the S/N:

$$L_{rcomp} = (\text{Switches, circulators, filters, etc.})$$

This loss, as was the case in the transmitter section, is a direct attenuation of the signal, 1 dB loss equals 1 dB loss in link margin.

1.6.5 Received Signal Power at the Input to the LNA

The received signal level at the input to the LNA is calculated as follows:

$$P_s = EIRP + A_{fs} + L_p + L_{ra} + G_r + L_{multi} + L_{rll} + L_{rcomp} \quad 1.11$$

This equation makes the assumption that all the losses are negative and the gain and EIRP are positive. Also, the above equation assumes that all the parameters are in either dB, dBm or dBw. Most often dBm is used as the standard method of specifying power. The signal power at input to the LNA is important to obtain an understanding of the S/N ratio for the receiver since the noise figure is generally established at this time by the LNA. The signal power can be improved by the process gain of some spread spectrum systems, for example a matched filter will increase the power by the amount of process gain. However, in most systems this equation establishes the signal part of the S/N.

Transceiver Design

The noise is determined by using the narrowest bandwidth of the system. Therefore, a preliminary S/N can be calculated to get an estimation of the performance of a receiver. The S/N can be calculated as follows:

$$S/N = (P_s + PG)/kTBF$$

where:

S/N = Signal to Noise ratio
P_s = power to the LNA
PG = process gain (matched filter)
K = Boltzman's Constant
T = Nominal Temperature (290 degrees)
B = Bandwidth
F = Noise Figure of the Receiver

1.6.6 Receiver Implementation Loss

When implementing a receiver in hardware, there are several components that do not behave ideally and there are losses associated with these devices that degrade the S/N. These need to be accounted for in the link budget. One of the contributors to this loss is the phase noise or jitter of all of the oscillator sources in the system including the up and down conversion oscillators, the code and carrier tracking oscillators, match filter and A/D oscillators, etc. Another source of implementation loss is the detector process including non-ideal components and quantization errors. This loss directly affects the receiver's performance and degrades the link budget directly. The losses are included as below:

L_i = implementation loss (departure from ideal, detector implementation, phase tracking of phase lock loops, etc.).

A ballpark figure for implementation losses for a typical receiver is about –4 dB. With more stable oscillators and close attention during receiver design, this can be reduced. This loss will vary from system to system and should be analyzed for each receiver system. This loss is a direct attenuation of the signal, one dB loss equals one dB loss in link margin.

1.6.7 Receiver Spreading Losses

The spreading losses are associated with non-ideal spreading and despreading techniques which reduce the signal power (sometimes referred to as match filter loss). This is an additional loss separate from the implementation loss since it deals with the match filter directly. Also since many digital systems are not spread spectrum systems the loss is kept separate for clarity. An example of this type of loss is the degradation realized when summing the pulses in a matched filter that has errors in the delays of the tapped delay line. The spread spectrum loss is generally around –1 to –2 dB and varies from system to system. Note that this is not included in a non-spread spectrum system. The losses are specified as follows:

Transceiver Design

L_{ss} = Spread Spectrum loss (−1 to −2 dB)

The spread spectrum loss not only affects the link budget but also affects the ability of the system to reject jammers (jamming margin). The value of the spread spectrum loss is dependent on the type of spread spectrum used and the receiver design for spreading and despreading. This loss is a direct attenuation of the signal, one dB loss equals one dB loss in link margin.

1.6.8 Process Gain

Process gain is the advantage over interfering signals received by despreading a spread spectrum signal.

G_p = process gain.

The process gain minus the spreading losses and the implementation losses provides the jamming margin for the receiver. The jamming margin is the amount of extra power, referenced to the desired signal, that the jammer must transmit in order to jam the receiver.

$$J_m = G_p + L_i + L_{ss} = \text{jamming margin.} \qquad 1.12$$

One question that is often times asked is "does spreading the signal increase the S/N of the system so that it requires less power to transmit." This depends on the definition used. In general, the spreading process does not improve S/N for a system. Sometimes this can be a bit confusing when only looking at the receiver. The receiver improves the S/N related to the input signal. This improvement is included in the link budget. The reason for this is because the bandwidth required for the spread signal is much larger on the input to the receiver than is needed for a non-spread system. Since the bandwidth is large on the input to the receiver, this bandwidth is either reduced to reduce the noise power, which produces a better S/N, or the signals are summed together producing a larger amplitude which produces a better S/N. However, if spread spectrum is not used, the bandwidth is much narrower at the input to the receiver which produces the same S/N and the despread system. Therefore, the S/N required at the transmitter for a non-spread system is the same as that required for a spread system. In actual spread spectrum systems, the S/N is degraded due to the spreading losses associated with the spreading and de-spreading processes.

One exception to the rule is a pulsed system with a specified pulse width and a given S/N. If this pulse is sent out multiple times instead of just once in a period and these pulses had pseudo-random phases and were combined in a matched filter, the signal would increase given the same noise bandwidth. Therefore, the system S/N would be improved. This is a very specific definition. If only one wider pulse without spread spectrum is used with a width equal to the length of the pseudo-random pulses, then there will not be improvement in S/N for the system. This is because the bandwidth is reduced which reduces the noise power by the same amount as the combined signal energy of the spread spectrum pulses.

Transceiver Design

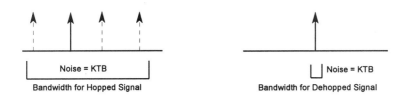

FREQUENCY DEHOPPING
S/N is Improved by Reducing the Bandwidth thus Reducing the Noise

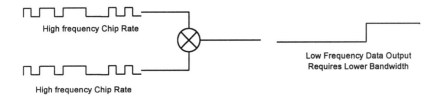

SLIDING CORRELATOR
S/N is Improved by Reducing the Bandwidth thus Reducing the Noise

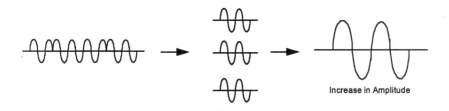

ANALOG MATCHED FILTER
S/N is Improved by Using a Delay line and Summing the Signal Sections for Increase Signal Output

Figure 1-5 *Methods of improving signal to noise for different spread spectrum techniques.*

There are slight differences in the process gain when referring to the link budget and when referring to the jamming margin. For example, in a frequency hopped system, the process gain referring to jamming is directly proportional to the number of frequencies. For process gain referring to the S/N improvement of the receiver, it is entirely dependent on the reduction of the bandwidth. Even in the jamming margin process gain example, the separation of the frequencies used can alter the actual jamming margin. For example, if the frequencies are too close together, the jammer may jam multiple frequencies at a time.

There are basically two ways to incorporate the processing gain in a spread spectrum receiver with regards to the link budget and jamming margin.

One of the ways to increase the S/N in the receiver is by reducing the bandwidth, thereby reducing the noise. For example, a frequency hopped signal has a constant amplitude but is changing frequency over a large bandwidth. The dehopper follows the hopped signal and the resultant bandwidth is greatly reduced so that the noise is reduced by the same amount, see Figure 1-5. However, the signal level is not increased, just the bandwidth is reduced. The narrow band jammer, however, is spread out by the dehopped process so that the amplitude of the jammer is lower than the desired signal in the narrow bandwidth which gives the process gain. The process gain (PG) for a frequency hopped waveform is defined as:

Transceiver Design

$$PG = 10\log(\text{number of frequency channels}) \quad 1.13$$

This is true for a narrowband intercept receiver that only receives one of the frequencies at a time, and for a narrowband jammer that only jams one of the frequencies at a time.

Another example is a continuous binary phase shift keyed (BPSK) signal that uses a sliding correlator for despreading. The bandwidth is spread due to the phase shift rate (chip rate) and the sliding correlator multiplies the incoming signal with the chip code which strips off the code and leaves the data rate bit stream which requires less bandwidth to process (see Figure 1-5). The narrow bandwidth reduces the noise level but the signal level is not increased. The jamming signal is spread out by the sliding correlator so that the amplitude of the jammer is lower than the desired signal in the narrow bandwidth, providing the process gain. The process gain is defined as:

$$PG = 10\log(\text{RF bandwidth/Detected bandwidth}) \quad 1.14$$

With this type of spread spectrum, the process gain for both the link budget and jamming margin is the same.

Another way to increase the S/N in the receiver is by summing the power in each bit of energy to increase the signal strength. This is accomplished by using a match filter similar to an Acoustic Charge Transport device (ACT) or a Surface Acoustic Wave device (SAW) (see Figure 1-5). This can also be done digitally by using a Finite Impulse

Response (FIR) filter with the tap spacing equal to the chip time with the weights either +1 or −1 depending on the code pattern. This system produces pulses when the code is lined up and data can be extracted using pulse position modulation (PPM). This does nothing to the bandwidth so the noise is not decreased, but it does integrate the signal up to a larger level. The process gain is 10log(#bits combined). The jammer is not coherent to the matched filter so it only produces sidelobes at the time of the system pulses and in between the system pulses. These sidelobe levels are reduced due to the length of the matched filter. This spread spectrum process also has the same definition for the process gain with regards to both link budget and jamming margin.

This last approach is used for low probability of intercept (LPI) system since the power can be integrated up to a larger pulse amplitude relative to a standard pulsed system. The length of the pulse determines the signal amplitude.

1.6.9 Received Power for Establishing S/N of a System

The detected power that is used to calculate the final S/N calculation to be used in the analysis is the following:

$$P_d = P_s + L_i + L_{ss} + G_p + G_{receiver} = S(dB) \qquad 1.15$$

∗ where $G_{receiver}$ is equal to all the receiver gain. This gain
∗ will be applied to both the signal and the noise and will be

cancelled out during the S/N calculation. Therefore, carrying the receiver gain through the rest of the analysis is optional. This equation applies only if the receiver uses power summation in the process as in the matched filter correlator. The sliding correlator and the frequency dehopper do not include the G_p term in the received power level since it is used in the bandwidth reduction to reduce the noise power. Therefore, if the signal increases in the process to obtain a better S/N, then the G_p term applies. If only the bandwidth is reduced through the process to obtain a better S/N, then the G_p term does not apply.

1.6.10 Received Noise Power

The noise power (N) of the receiver is compared to the signal power (S) of the receiver to determine a power S/N ratio. The noise power is initially specified using temperature and bandwidth. System induced noise is added to this basic noise to establish the noise power for the system.

1.6.10.1 Noise Figure

The standard noise equation for calculating the noise at the receiver is:

$$N = kT_oBF \qquad 1.16$$

where:

T_o = nominal temp (290 degrees K).
F = Noise factor
k = Boltzman's constant
B = Bandwidth

The noise figure is the noise factor in dB or 10logN. The noise factor is used as a multiplier and the noise figure is in dB which is additive. The noise factor or figure is the change in S/N_{out} compared to the S/N_{in}. Noise factor is the ratio of the S/N_{out} and S/N_{in} and noise figure is the difference between them in dB.

The noise will be kT_oBF plus any affects due to the difference in temperature with F being the receiver noise factor as shown:

$$N = kT_oBF + (kT_sB - kT_oB) = kT_oBF + kB(T_s-T_o)$$
$$= kT_oBF_t \qquad 1.17$$

Solving for F_t:

$$F_t = F + (T_s-T_o)/T_o \qquad 1.18$$

where:

T_s = sky temp (52 degrees).

If $T_s = T_o$ then the equation 1.16 applies.

Since generally T_s is less than T_o, the noise will be less than the standard kT_oBF and the noise factor is reduced by $(T_s-T_o)/T_o$.

Transceiver Design

The noise figure equals the LNA noise figure, F_{LNA}, assuming that the losses are not too great compared to the gain of the receiver. If the losses plus the next amplifier's noise figure are high relative to the gain of the first amplifier, then the noise factor is calculated by:

$$F_t = F_1 + [(F_2 + \text{Losses})-1]/G_1. \qquad 1.19$$

The complete noise factor equation is:

$$F_t = F_1 + [(F_2+\text{Losses})-1]/G_1 + [(F_3+\text{Losses})-1]/G_1 G_2 \,.. \qquad 1.20$$

The noise figure is the LNA (and all additional losses from the following stages according to the equation 1.20) since the losses in the preceding stages are included in the budget. If the noise figure is calculated at the antenna, then the line losses would not be included in the link budget but would affect the noise figure directly, one dB loss would be one dB higher noise figure. The reason for this is that the noise temperature is the same before and after the losses, so the kT noise is the same but the signal is attenuated. Therefore, the S/N is reduced on the output. There is no additional noise added to the system, however, noise figure has been defined as $S/N_{in}-S/N_{out}$ (in dB). This is true only if the bandwidths of the input and the output are the same. One major assumption is that the bandwidth is smaller as the signal progresses through the receiver. If the bandwidth increases, then the noise figure increases. This noise figure increase could be substantial and may need to be included in the analysis.

1.6.10.2 Received Noise Power at the Detector

The noise power at the detector is computed converting the paramenters of equation 1.14 into dB to simplify the calculation and including the gain of the receiver $G_{receiver}$ the received noise power at the detector is:

$$N(dB) = kT_o(dB) + B(dB) + F_{LNA}(dB) + G_{receiver} \quad 1.21$$

Note that this assumes $T_s = T_o$ and that the line losses are included in the link budget and not the noise figure. Eliminating $G_{receiver}$ for analysis purposes is again optional since this is common to both the noise power and the signal power.

1.6.11 Receiver Bandwidth

The receiver bandwidth often used in determining the final S/N is equal to the bit rate for BPSK. The energy in a bit, E_b, times the bit rate is equal to the signal power S for BPSK, and the noise power spectral density N_o (which is the amount of noise per hertz, or in other words the amount of noise in a 1 Hz bandwidth), times the bandwidth gives the total noise power N. Therefore, if the bandwidth is equal to the bit rate, then S/N = E_b/N_o.

The bandwidth used in the link budget is the bit rate bandwidth. This provides an approximation so that S/N can approximate E_b/N_o since S/N is easily measured and E_b/N_o is used to find the probability of error from the curves [4]. The curves are generated using the

Transceiver Design

mathematical error functions, erf(x). For analysis purposes, QPSK would be the same as BPSK when the bit rate bandwidth is used. However, when actually measuring the S/N for QPSK, the measurement of the noise would be the symbol rate bandwidth which is 1/2 the bit rate bandwidth, since the symbol rate is 1/2 the bit rate. Therefore, using QPSK, the bit rate is twice as fast as the BPSK rate since twice as many bits are sent in the same noise bandwidth, assuming equal bandwidths. The noise bandwidth is calculated using the symbol rate bandwidth, and the bit rate is twice the symbol rate.

Therefore:

$$S/N = (E_s/N_o)(\text{symbol rate})/(\text{symbol rate BW}). \quad 1.22$$

$$E_s = 2E_b \quad 1.23$$

Thus:

$$S/N = 2E_b/N_o \quad 1.24$$

so:

$$E_b/N_o = .5 S/N \quad 1.25$$

Note that the S above contains twice the power as it would have if just the bit rate signal was measured. Therefore, if the analysis uses the bit rate bandwidth for the noise power and the bit rate, then the $S/N = E_b/N_o$. QPSK can be viewed as either sending out twice as many bits in the same bandwidth as BPSK, or sending out the same amount of bits in half the bandwidth.

1.6.12 Received S/N at the Detector

The signal to noise is thus calculated by subtracting the noise power in dB from the signal power in dB:

$$S/N(dB) = S(dB) - N(dB) \qquad 1.26$$

This is the S/N used for the analysis in determining the performance of the detector.

1.6.13 Received E_b/N_o at the Detector

The received E_b/N_o(dB) is equal to the received S/N(dB) for BPSK for an unspread signal. For a spread spectrum signal, the bandwidth will be larger than the bit rate, so the S/N(dB) = E_b/N_o(dB) – G_p.

For other waveforms, the equation varies on the type of modulation of the signal used in the calculation of S/N. For QPSK, if the signal is the bit rate measurement, then the same equation holds. If the symbol rate is used, then the factor of 0.5 applies. The E_b/N_o is the critical paramenter for measurement of the performance of the system. Once this has been determined, it is then compared to the required E_b/N_o to determine if it is adequate for proper system operation for a given probability of error.

Transceiver Design

1.6.14 Required Uncoded Ideal E_b/N_o

The required uncoded ideal E_b/N_o is the E_b/N_o that is necessary to achieve the specified probability of error. This is calculated using error functions and is usually obtained from probability curves. For example,

$$P_o = 1 \times 10^{-8} \text{ for } E_b/N_o = 12 \text{ dB} \qquad 1.27$$

where:

P_o = probability of error

E_b/N_o for an uncoded coherent BPSK

This gives the probability of bit error for an ideal system. If this is too low, then the E_b/N_o needs to be increased, usually by increasing the power out of the transmitter. This also can be done by changing any of the link budget parameters discussed previously, for example, reducing the line loss.

1.6.15 Receiver Coding Gain

The coding gain is the gain achieved by implementing certain codes to improve the BER thus requiring a smaller E_b/N_o for a given probability of error:

G_c = coding gain

The amount of gain depends on the coding scheme used.

For example, a Reed-Solomon code rate 7/8 gives approximately 4.5 dB coding gain at a BER of 10^{-8}.
Note that coding requires a tradeoff since adding the coding either increases the bandwidth by 8/7 or the bit rate is lowered by 7/8.

1.6.16 E_b/N_o Required

The required E_b/N_o for the link budget is:

$$E_b/N_o(\text{req.}) = E_b/N_o(\text{uncoded}) - G_c \qquad 1.28$$

This is the minimum required E_b/N_o to enable the transceiver to operate correctly. The link budget parameters need to be set to ensure that this minimum E_b/N_o is present at the detector.

1.6.17 Link Margin

Since there are both known and unknown variances in the link budget, a margin is established to ensure proper system operation. The Link margin (additional E_b/N_o over the required E_b/N_o) for the system is:

$$\text{Link margin} = E_b/N_o - E_b/N_o(\text{req.}) \qquad 1.29$$

Transceiver Design

1.6.18 Link Budget Example

A typical link budget spreadsheet is shown in Table 1-1 which is used for calculating the link budget. This is done in Excel and starts out with the slant range, frequency, and power output. This link budget tracks the power level and the noise level side by side so that the S/N can be calculated anywhere in the receiver. The altitude of an aircraft or satellite and its angle incident to the earth is used to compute the slant range and the atmospheric loss is analyzed at the incident angle. The bit-rate bandwidth plus the expansion due to the code is used to determine the final noise bandwidth.

1.7 Summary

The link budget provides a means to design a transceiver and to perform the necessary analysis to ensure that the design is optimal for the given set of requirements. The link budget provides a way to easily make system tradeoffs and to ensure that the transceiver operates properly over the required range. Known or specified parameters are entered, and the link budget solves for the unknown parameters. If there is more than one unknown parameter, then the tradeoffs need to be considered. The link budget is continually re-worked, including the tradeoffs to obtain the best system design for the link. Generally a spreadsheet is used for ease of changing the values of the parameters and seeing the effects of the change (see in Table 1-1).

Transceiver System Design

Table 1-1 *Typical link budget analysis.*

Link Budget Analysis				
	Slant Rng	Freq.	Power	
Enter Constants.........	92.65	3	1	
Enter Inputs............	Inputs	Power Levels		
Transmitter	Gain/Loss	Sig.(dBm)	Noise(dBm)	
Tran.Pwr(dBm)=		30		
Trans. line loss =	-1	29		
Other(switches)	-2	27		
Trans Ant Gain =	10	37		
Ant. Losses=*	-2	35		
ERP		35		
Channel				
Free Space Loss =	-141.32	-106.32		
Rain Loss =	-2	-108.32		
Cloud Loss =	-1	-109.32		
Atm loss(etc) =	-0.5	-109.82		
Multipath Loss=	-2	-111.82		
Receiver				
RF BW(MHz)**	100		-94	
Ant. losses =*	-2	-113.82		
Rx Ant Gain =	10	-103.82		
Ant. ohmic loss	-2	-105.82		
Other(switches)	-1	-106.82		
Rec. Line loss =	-2	-108.82		
LNA Noise Fig. =	3		-91	
LNA Gain =	25	-83.82	-66	
NF Deviation =	0.1		-65.9	
LNA levels =		-83.82	-65.9	
Receiver Gain =	60	-23.82	-5.9	
Imp. Loss	-4	-27.82		
Despread BW**	10		-15.9	
PG 10log# of bits	30	2.18		
Spreading Loss =	-2	0.18		
Match Filt. loss				
Res. Doppler				
Det. Peak offset				
Det. Levels =		0.18	-15.9	
S/N =		16.08		
Req. Eb/No				
Req. Eb/No		12	Pe=10exp-8	10^{-8}
Coding Gain =	4	8		
Eb/No Margin =		8.08		

1.8 References

[1] Om P. Gandhi, *Microwave Engineering and Applications*, 1981, Pergamon Press Inc.

[2] Robert K. Crane, *IEEE Transactions on Communications*, Vol. Com-28, No. 9, September 1980.

[3] Mischa Schwartz, *Information, Transmission, Modulation and Noise*, 1980, McGraw-Hill.

[4] Simon Haykin, *Communication Systems*, John Wiley & Sons Inc., New York, 1983.

Problems

1. Show that a specification of:

   0 dBm ± 2 dBm =

   is an impossible statement and write a correct statement for it.

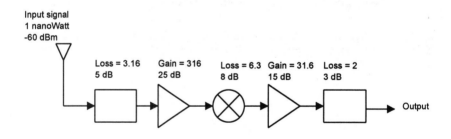

2. Solve for the total signal level through the receiver using milliwatts to obtain the answer, and then solve for the output in dBm for the above receiver.

Transceiver Design

Compare your answers. Note: Try the dB method without using a calculator.

3. An oscillator generates a center frequency of 1 GHz at 1 mW. The largest spurious response, or unwanted signal is at a power level of .01 mW. How would you specify the spurious response in dBc? What would the power be in milliwatts for a spur at −40dBc?

4. A system has been operational for the past 3 years. A need arose to place a limiter in the path between the antenna and the LNA to avoid overload from a newly built high power transmitter. The limiter has 1.5 dB of loss. How will this affect the ability to receive low level signals. What could you do to overcome the loss?

5. What is the diameter of a parabolic antenna operating at 5 GHz, at an efficiency of .5 and a gain of 10 dB?

6. What is the freespace attenuation for the system in problem 5 with a range of 10 nautical miles?

7. What is the noise level at the LNA given that the bandwidth is 10 MHz and the LNA noise figure is 3 dB with $T_o = T_s$ at room temperature?

8. What is the noise figure of the receiver, given that the LNA has a noise figure of 3 dB, a gain of 10 dB,

48   Transceiver System Design

a second amplifier after the LNA, with a gain of 20 dB, a noise figure of 10 dB, and there is a loss of 5 dB between the amplifiers?

9. What is the null-to-null bandwidth of a BPSK signal running at a chip rate of 20 MHz?

10. What is the noise level of problem 2 if the system noise figure is 3 dB at the LNA and a bandwidth of 10 MHz, and what is the output signal-to-noise S/N in dB?

2

The Transmitter

The transmitter is responsible for formatting, encoding, modulating and upconverting the data that is transmitted over the link with the required power output according to the link budget analysis. The transmitter section is also responsible for spreading the signal using various spread spectrum techniques. Several spread spectrum waveforms are discussed in this chapter. The primary types of spread spectrum using direct sequence methods to phase modulate a carrier are discussed in detail including diagrams and possible design solutions. A block diagram showing the basic components of a typical transmitter is shown in Figure 2-1.

2.1 Antenna

The antenna receives the RF signal from the power amplifier and translates this RF power into electromagnetic waves so that the signal can be propagated through the air. The antenna is dependent on the frequency of the RF and the specified operational requirements. The proper design of the antenna ensures that the maximum signal power is sent out in the direction of the receiver or in the required service volume. The design of the antenna is not a trivial task and some engineers devote their lives in trying to design a better, more efficient antenna. The gain of the antenna improves the link budget on a one for one basis. That is, for every dB of gain the antenna exhibits, the link is improved by a dB. Therefore, careful design of the antenna can reduce the power output required from the power amplifier which

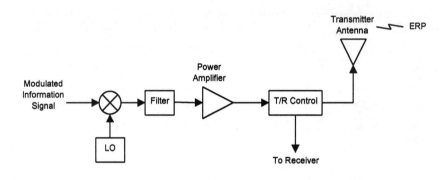

Figure 2-1 Block diagram of a basic transmitter.

reduces cost of the system. The frequency, amount of gain required, and size, are factors in determining the type of antenna to be used. Parabolic dishes are used frequently at microwave frequencies. The gain of the parabolic dish was calculated in the link budget section of Chapter 1. The antenna provides gain in the direction of the beam to reduce power requirements and also to reduce the amount of interference received which is called selective jamming reduction. The gain of the antenna is usually expressed in dBi which is the gain in dB reference to what an isotropic radiator antenna would produce. In other words, this is the amount of amplifier gain that would be delivered to an isotropic radiator antenna to transmit the same amount of power in the same direction as a directional antenna. Even a dipole antenna provides gain in the direction of the receiver, which is typically 2.1 dBi.

2.2 Transmit/Receive Control

In order to reduce the number of antennas for a transceiver system which ultimately reduces cost, the same antenna is generally used for both the transmitter and the receiver at each location. If the same antenna is going to be used for the system, a device is required to prevent the transmitter from interfering with or damaging the receiver. This device can be a duplexer or diplexer, T/R switch, circulator or some combination of these devices. The method chosen must provide the necessary isolation between the transmitter circuitry and the receiver circuitry and prevent damage to the receiver during transmission.

The transmitted signal is passed through this device and out to the antenna during the time to transmit with adequate isolation from the receiver. The received signal is then passed through the device to the receiver with adequate isolation from the transmitter.

2.3 Upconversion

The information that is to be transmitted is made up of low frequencies, such as voice or a digital data stream. The low frequencies are mixed or upconverted to higher frequencies before radiating the signal out of the antenna. One reason for using higher frequencies for radiation is the size of the antennas. For example, many antenna designs require element lengths of $\lambda/2$ where λ becomes rather large with low frequencies ($\lambda = 1/f$). Using a frequency of 10 kHz, a element length of $\lambda/2$ would equal 15,000 meters. Using a frequency of 1 GHz, a element length of $\lambda/2$ would equal .15 meters. Another reason is limited bandwidth. The lower usable frequency bands are overcrowded and therefore the operating frequencies continue to increase in hopes that there will be a less crowded band to operate in. Regardless of the reason, the bottom line is that one of the main functions of the transmitter is to upconvert the signal to a usable radiating frequency.

A local oscillator (LO) is used to translate the lower frequency band to a higher frequency band for transmission. For example, a single lower frequency is mixed or multiplied by a higher frequency which results in the sum and difference of the two frequencies as shown:

The Transmitter

$$\cos(\omega_c t+\phi_c)\cos(\omega_l t+\phi_l)=\frac{1}{2}\cos[(\omega_c-\omega_l)t+\phi_c-\phi_l]$$
$$+\frac{1}{2}\cos[(\omega_c+\omega_l)t+\phi_c+\phi_l] \qquad 2.1$$

where:

ω_c = carrier or higher frequency
ϕ_c = phase of the carrier
ω_l = the lower frequency containing the data
ϕ_l = phase of the lower frequency
t = time

This is an ideal case that ignores any of the harmonics of the two input frequencies and only uses one low frequency (usually the data includes a band of frequencies). A band pass filter is used at the output to select either the sum term or the difference term above. The sum term is selected in this example. A band pass filter is centered at $\cos[(\omega_c+\omega_1)t+\phi_c+\phi_1]$ with a roll off sharp enough to virtually eliminate the term $\cos[(\omega_c-\omega_1)t+\phi_c-\phi_1]$ with the result shown below:

$$\frac{1}{2}\cos[(\omega_c+\omega_l)t+\phi_s] \qquad 2.2$$

where:

ϕ_s = sum of the phases
$\omega_c+\omega_l$ = sum of the frequencies

This signal is then sent out very easily at the frequency of choice.

2.4 LO and Elimination of Sideband

One of the sidebands must be filtered out in order to recover the signal or a different downconverting frequency needs to be used. If neither of the above conditions are met, there is a possibility that the phases will align to eliminate the signal for a particular phase. For example, suppose that the signal to be transmitted is defined as:

$$A\cos(\omega_s t)$$

and the carrier frequency is defined as:

$$B\cos(\omega_c t)$$

The output of the upconversion process is found by multiplying the two signals as shown:

$$A\cos(\omega_s)t \; B\cos(\omega_c)t \\ = \frac{AB}{2}\cos(\omega_s+\omega_c)t + \frac{AB}{2}\cos(\omega_s-\omega_c)t \qquad 2.3$$

Generally, one of the above terms are filtered out. Now, assume the worst case phase situation; mixing down with

The Transmitter

the LO 90 degrees out of phase $(\cos(\omega_c t + 90) = -\sin(\omega_c t))$ with no filtering. The results are:

$\cos(\omega_c t - 90) = \sin\omega_c t$

$$[\frac{AB}{2}\cos(\omega_s+\omega_c)t + \frac{AB}{2}\cos(\omega_s-\omega_c)t][\sin(\omega_c t)] \qquad 2.4$$

$$= \frac{ABC}{4}[\sin(\omega_s+\omega_c+\omega_c)t + \sin(\omega_s-\omega_c+\omega_c)t]$$

$$- \frac{ABC}{4}[\sin(\omega_s+\omega_c-\omega_c)t + \sin(\omega_s-\omega_c-\omega_c)t]$$

$$= \frac{ABC}{4}[\sin(\omega_s+2\omega_c)t + \sin(\omega_s)t]$$

$$- \frac{ABC}{4}[\sin(\omega_s)t + \sin(\omega_s-2\omega_c)t]$$

$$= \frac{ABC}{4}[\sin(\omega_s+2\omega_c)t - \sin(\omega_s-2\omega_c)t]$$

The results show that only the sum of $\omega_s + 2\omega_c$ and the difference of $\omega_s - 2\omega_c$ frequencies were the outputs of this process. The frequency ω_s was not retrieved with this particular phase situation. This is the worst case, but the point is that it can happen and will degrade the output substantially. This demonstrates the requirement to filter out one of the unwanted sidebands.

If one of the sidebands were filtered, (for example the upper sideband), then the results would be:

$$[\frac{AB}{2}\cos(\omega_s-\omega_c)t][\sin(\omega_c t)] \qquad 2.5$$

$$= \frac{ABC}{4}[\sin(\omega_s-\omega_c+\omega_c)t] - \frac{ABC}{4}[\sin(\omega_s-\omega_c-\omega_c)t]$$

$$= \frac{ABC}{4}[\sin(\omega_s)t] - \frac{ABC}{4}[\sin(\omega_s-2\omega_c)t]$$

The results of filtering the upper sideband with this worst case scenario shows that the signal waveform ω_s is retrieved along with the difference of $\omega_s - 2\omega_c$. A low pass filter will eliminate the second term since the carrier frequency is generally much higher than the signal frequency (actually 2 times the carrier frequency) giving only the desired signal output of {ABC/4[sin(ω_s)t]}.

Most transmitters will filter out one of the upconversion products from the transmitted signal to eliminate this problem and also to reduce the amount of power output required from the power amplifier.

Another way to eliminate the sideband in the transmitter is to use a image-reject mixer. The image-reject mixer quadrature upconverts the signal into I and Q and then phase shifts the I channel by −90 degrees to eliminate the unwanted sideband.

First of all the quadrature upconversion produces the following:

The Transmitter

$$I \text{ Channel Output} = \cos(\omega_s)\cos(\omega_c) \tag{2.6}$$
$$= \frac{1}{2}\cos(\omega_s-\omega_c)+\frac{1}{2}\cos(\omega_s+\omega_c)$$

$$Q \text{ Channel Output} = \cos(\omega_s)\cos(\omega_c-90) \tag{2.7}$$
[handwritten: $\sin \omega_c$]
$$= \frac{1}{2}\cos(\omega_s-\omega_c+90)+\frac{1}{2}\cos(\omega_s+\omega_c-90)$$

The I channel output is then phase shifted by −90 degrees and the I and Q channels are then summed together to produce the following:

$$I \text{ Channel Phase Shifted} = \frac{1}{2}\cos(\omega_s-\omega_c-90) \tag{2.8}$$
$$+\frac{1}{2}\cos(\omega_s+\omega_c-90)$$

$$I + Q = \frac{1}{2}\cos(\omega_s-\omega_c-90)+\frac{1}{2}\cos(\omega_s+\omega_c-90) \tag{2.9}$$
$$+\frac{1}{2}\cos(\omega_s-\omega_c+90)+\frac{1}{2}\cos(\omega_s+\omega_c-90)$$
$$= \cos(\omega_s+\omega_c-90)$$

Therefore, only the upper sideband is transmitted, thus eliminating the problem on the receive side.

Another possibility for eliminating the problem is to use a quadrature downconversion on the receiver side. The transmitter remains the same and transmits both sidebands as shown below:

$$\text{Transmitter Output} = \cos(\omega_s)\cos(\omega_c)$$
$$= \frac{1}{2}\cos(\omega_s-\omega_c)+\frac{1}{2}\cos(\omega_s+\omega_c)$$

2.10

The receiver processes the incoming signal using a quadrature downconverter so that when the worst case situation occurs with the $\sin(\omega_c)$ multiplying, the quadrature channel will be multiplying with a $\cos(\omega_c)$ so that the signal is recovered in either (or both) the I and Q channels as shown:

$$I\ Channel = [\frac{AB}{2}\cos(\omega_s+\omega_c)t$$
$$+\frac{AB}{2}\cos(\omega_s-\omega_c)t][\sin(\omega_c t)]$$
$$= \frac{ABC}{4}[\sin(\omega_s+2\omega_c)t-\sin(\omega_s-2\omega_c)t]$$

2.11

$$Q\ Channel = [\frac{AB}{2}\cos(\omega_s+\omega_c)t$$
$$+\frac{AB}{2}\cos(\omega_s-\omega_c)t][\cos(\omega_c t)]$$
$$= \frac{AB}{2}\cos(\omega_s)t+\frac{AB}{4}\cos(\omega_s+2\omega_c)t$$
$$+\frac{AB}{4}\cos(\omega_s-2\omega_c)t$$

2.12

The Transmitter

This shows all of the signal being received in Q channel, $\cos(\omega_s)$, and none in the I channel. Most of the time this will be a split, and the magnitude and phase will be determined when combining both the I and Q channels. This quadrature technique uses carrier recovery loops when receiving a suppressed carrier waveform. For example, Costas Loops, which are commonly used in direct sequence systems, use this technique to recover the carrier and down convert the waveform into I and Q data streams.

2.5 Power Amplifier

The power amplifier is used to provide the necessary power output of the transmitter to satisfy the link budget for the receiver. The general classifications for the power amplifier that are commonly used are:

Class A: Power is on all of the time (linear operation).

Class B: Power is on for 1/2 of the time for one stage and 1/2 of the time for the other stage, usually in a push-pull configuration. That is, one stage is on and the other stage is off.

Class C: Power is on for less than 1/2 of the time (non-linear operation).

There are also various combinations of the above classes not mentioned (Class AB, etc.).

The antenna is matched to the same impedance as the power amplifier to deliver maximum power to the load. Generally this is specified using a 50 ohm system and the voltage standing wave ratio (VSWR) is the measurement used to indicate the amount of impedance mismatch. The power amplifier is often the variable in the link budget. The slant range is generally specified and the power amplifier is chosen to deliver enough power to meet the link budget. The power amplifier is generally a high cost item in the transceiver. Therefore, careful transceiver design to minimize the losses in the link budget will reduce the power requirement of the power amplifier thus reducing the overall cost of the tranceiver.

2.6 VSWR

The voltage standing wave ratio (VSWR) defines a standing wave that is generated on the line between the power amplifier and the antenna. The ratio of the minimum and maximum voltages measured along this line is the VSWR. For example, a VSWR of 2:1 defines a mismatch where the maximum voltage is twice the minimum voltage. The first number defines the maximum voltage and the second is the normalized minimum. The minimum is defined as one and the maximum is the ratio of the two.

The Transmitter   61

The standing wave is caused by reflections of the signal, due to a mismatch in the load, that are returned along the transmission line. The reflections are added by superposition causing destructive and constructive interference with the incidence wave depending on the phase relationship between the incidence wave and the reflected wave. This results in voltages that are larger and smaller on points along the line. The main problem with mismatch and a high VSWR is that some of the power is lost when it is reflected back to the source and is not delivered to the load or antenna.

The standing wave is minimized by making the impedances equal so that there are virtually no reflections, which produces a VSWR of 1:1.

2.7 Spread Spectrum Transmitter

Many systems today use spread spectrum techniques to transport data from the transmitter to the receiver. One of the most common forms of spread spectrum uses phase-shift keying and is referred to as Direct Sequence (DS). The data is usually exclusive-or'd with a pseudo-random or pseudo-noise (PN) code that has a much higher chipping rate than the data rate. This produces a much wider occupied spectrum in the frequency domain (spread spectrum).

DS systems use phase shift generators (PSG) to transfer data (plus code for spread spectrum systems) by phase shifting a carrier frequency. There are several ways to build a PSG depending on the waveform selected. Detailed description of phase-shift keying modulation is provided.

Other forms of spread spectrum including frequency hopping, time hopping, chirped FM, and ways to manage multiple users are included.

2.7.1 Phase-Shift Keying

Phase-shift keying (PSK) is a type of modulation where the phase of the carrier is shifted to different phase states by discrete steps using a digital sequence. This digital sequence can be either the digitized data or a combination of digitized data and a spread spectrum sequence. There are many different levels and types of PSK. This discussion will be limited to a maximum of four phase states. However, the principle can be extended to higher order PSK.

2.7.1.1 Binary Phase Shift Keying (BPSK)

The basic PSK is the binary-PSK or BPSK (see Figure 2-2). This is defined as shifting the carrier 0 or 180 degrees in phase depending on the digital waveform. For example, a +1 gives 0 degrees phase of the carrier, and −1 shifts the carrier by 180 degrees. To produce the digital waveform, the data or information signal is digitized, encoded, and sent out in a serial bit stream, (if not already), and modulo 2 (exclusive-or) added to a PN sequence. The end result is

The Transmitter

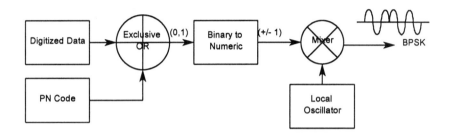

Figure 2-2 *BPSK generator.*

a serial modulating digital waveform. The output of the modulo 2 adder (exclusive-or) contains 0 and 1 and needs to be changed to ± 1 for the mixer to operate. However, certain forms of hardware can bypass this step and modulate the mixer directly. Emitter Coupled Logic (ECL) contains differential outputs and can be connected directly to the mixer. ECL is also capable of driving 50 ohms directly without an additional driver. A dual input mixer is required in order for the ECL logic to connect directly.

Applying minus voltage to the mixer reverses the current through the balun of the mixer and causes the current to flow in the opposite direction to create a net 180 degree phase shift of the carrier. Therefore, the carrier is phase shifted between 0 and 180 degrees depending on the input waveform. A simple way of generating BPSK is shown in Figure 2-2. The LO is either multiplied by a +1 or a −1 from the digital sequence producing a 0 or a 180 phase shift.

Other devices such as phase modulators or phase shifters can create the same waveform just as long as one digital level compared to the other digital level creates a 180 degrees phase difference in the carrier output.

2.7.1.2 Differential Phase Shift Keying (DPSK)

The BPSK waveform above can be sent out as absolute phase, that is a 0 degree phase shift is a "1" and a 180 degree phase shift is a "0". Another way to perform this function is to use differential PSK (DPSK) which monitors the change of phase. That is, a change of phase (0 to 180 or 180 to 0) represents a "1" and no change (0 to 0 or 180 to 180) represents a "0". This scheme is easier to detect because the absolute phase does not need to be determined, just the change of phase is monitored. Differential mode can be applied to various phase shifting schemes and higher order phase shift schemes. Differential results in about one dB of degradation compared to coherent PSK, however, it is dependent on the S/N level and the operating position on the probability of error curve. Note that differential can be applied in higher order PSK systems such as differential quaternary phase shift keying (DQPSK) and differential eight phase shift keying (D8PSK), etc.

The Transmitter

2.7.1.3 Quaternary Phase Shift Keying (QPSK)

The LO is quadrature phase shifted so that four phasors are produced on the outputs of the two mixers, 0 or 180 degrees out of one mixer and 90 or 270 degrees out of the second mixer. The phasor diagram shows the four phasors (see Figure 2-3). The two BPSK systems are summed together which gives four possible resultant phasors, 45, 135, 225, or 315 degrees which are all in quadrature as shown in Figure 2-3.

A quaternary-PSK or QPSK (sometimes called quadriphase PSK) can be done using two BPSK circuits in quadrature and then summing the outputs to generate four different phases. Since the digital transitions occur at the same time, changes between any four resultant phasors can occur. A phasor diagram of QPSK is shown in Figure 2-3.

One of the BPSK generators phase shift the carrier 0 and 180 degrees according to the standard BPSK generators. The second BPSK channel is identical except that the carrier is shifted by –90 degrees before being phase shifted by the bit stream. Therefore, the resultant phase shifts of the second BPSK channel will be 90 and 270 degrees. The summation on the phasor diagram produce four phase states at 45, 135, 225, and 315. Therefore, depending on the input of both bit streams, the phase of the resultant could be at any of the four possible phases. For example, if both bit streams were +1, then the phasor would be 45. If both bit streams changed to –1, then the phasor would be 225 degrees giving a change of carrier phase of 180

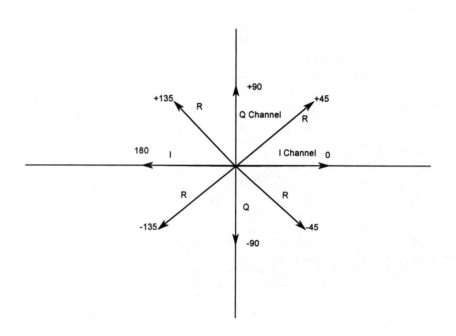

Figure 2-3 *QPSK phasor diagram using two BPSK modulators.*

degrees. If only the first channel changes to a +1, then the phasor would be at 315 degrees giving a change of 90 degrees. Therefore, considering all the possibilities, a ± 90, 0, 180 degree phase shifts are possible in this configuration.

The Transmitter

A simple way of generating QPSK is shown in Figure 2-4. Two different digital streams are used for each of the BPSK sections. The method of generation depends on the type of system used. The transitions of the two sequences occur at the same time in order to provide phase shifts of 0, 90, −90, and 180 degrees.

Usually the LO is phase shifted instead of the actual binary input. Either way would provide the phase shifts required. However, phase shifting the binary input requires phase shifting all of the frequency components of

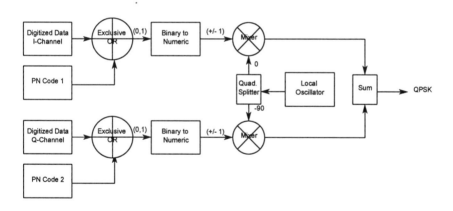

Figure 2-4 *QPSK generator.*

the digital waveform. This would require a more sophisticated phase shifter that is broadband. Shifting the carrier only requires a phase shift at one frequency, which is much simpler to build. In fact, in the latter case, a longer piece of cable cut to the right length can provide this phase shift.

2.7.1.4 Offset QPSK (OQPSK) or Staggered QPSK (SQPSK)

Another type of quadrature PSK is referred to as offset-QPSK (OQPSK) or staggered-QPSK (SQPSK). This configuration is identical to QPSK except that one of the digital sequences is delayed by a half cycle so that the phase shift only occurs on one mixer at a time. Therefore, the summation of the phasors can only result in a maximum phase shift of ±90 degrees and can only change to adjacent phasors (no 180 phase shift possible). The phasor diagram is identical to Figure 2-3, remembering that no 180 degree phase shifts are possible. This prevents zero crossover points and provides smoother transitions with less chance of error and reduces the amplitude modulation effects common to both BPSK and QPSK. The zero crossover points means that during the transition from 0 degrees to 180 degrees the amplitude of the phasor goes to 0. Therefore, since the hardware does not allow for an instantaneous change in phase, the changing amplitude produces amplitude modulation. The OQPSK only changes 90 degrees so that the phasor only goes through a −3dB amplitude degradation point. This reduces the amplitude modulation tremendously. Figure 2-5 shows a simple way

The Transmitter

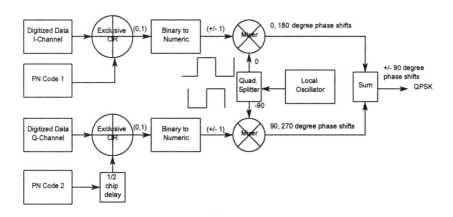

Figure 2-5 OQPSK generator.

of generating OQPSK. The only difference between this and the QPSK generator is the 1/2 chip delay in one of the bit streams. This prevents the 180 degree phase shift present in the QPSK waveform.

2.7.1.5 Higher Order PSK

The above analysis can be extended for higher order phase-shift keying systems. The same principles apply, only they are extended for additional phase states and phase shifts.

Direct Sequence waveforms are used extensively in the spread spectrum industry. BPSK is the simplest form of phase-shift keying and shifts the carrier 0 or 180 degrees. QPSK uses two BPSK systems, one in quadrature, to achieve four phase shifts, 0, ±90, and 180 degrees. OQPSK is uses to minimize amplitude modulation and is identical to QPSK with the phase transitions occurring at 1/2 chip intervals and at different times for each of the BPSK channels. This provides phase shifts of 0, and ±90 degrees and eliminates the 180 degree phase shift. Higher order PSK systems can be analyzed much the same way only with more phase states and phase transitions. However, the more phase shift possibilities, the harder it is to detect and resolve the different phase states. Therefore, there is a limit on how many phase states that can be sent for good detection. This limit seems to grow with better detection technology, but caution must be given to the practicality of how many phase states can be sent out for standard equipment.

2.7.1.6 Variations in PSK Schemes

There are many other PSK configurations that are in use today and there will be other types of modulating schemes in the future. One of the main concerns using PSK modulators is the amplitude modulation (AM) that is inherent when phase states are changed. For example, BPSK switches from 0 to 180 degrees. Since this is not instantaneous due to bandwidth constraints, the phasor passes through zero amplitude in the process. The more bandlimiting, the more amplitude modulation there is in

The Transmitter

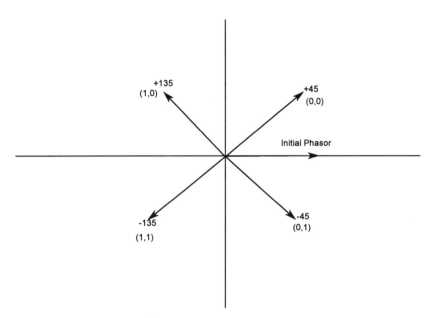

Figure 2-6 *Phasor diagram for a π/4 DPSK modulation waveform*

the resultant waveform. Various schemes have been developed to reduce this amplitude modulation problem. For example, OQPSK only allows a maximum phase shift of 90 degrees, so that the AM problem is significantly reduced. Also, MSK is used to smooth out the phase transitions which further reduces the AM problem and makes the transition continuous.

Other variations in PSK schemes are listed below which are some of the more commonly used ones.

2.7.1.7 π/4 Differential QPSK

The π/4 differential QPSK is a modulating scheme that encodes 2 bits of data as ±45 and ±135 degree phase shifts from the last symbol received. Because it is differential, once the phase shift occurred, the next phase shift is from the phase measured. Therefore, if the first phase shift was +135 degrees and the second phase shift was +135 degrees, the resultant phasor's absolute phase would be at 270 degrees. Therefore, the absolute phase could be 0, 45, 90, 135, 180, 225, 270, and 315 degrees. For example, if +45 degree shift occurred, then the bits that were sent was 0,0 (see Figure 2-6). Proceeding with the other phase shifts, a

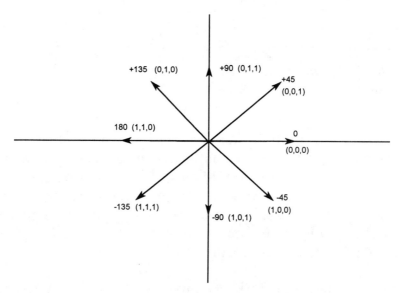

Figure 2-7 *Phasor diagram for D8PSK modulation waveform.*

The Transmitter

−45 degree shift is 0,1, a +135 degree shift is 1,0, and a −135 degree shift if 1,1 as shown in Figure 2-6. The shifts for the next bits sent start from the last phase measured. Note that the phase shift does not go through zero so that there is less AM for this type of modulation compared to DQPSK which does go through zero.

2.7.1.8 Differential 8PSK (D8PSK)

The D8PSK type of modulation is the same as $\pi/4$ DQPSK but includes the phase shift possibilities of 0, 90, −90, and 180 degrees thus providing 0, 45, −45, 90, −90, 135, −135, and 180. This provides 8 possible phase shifts or 3 bits of information as shown in Figure 2-7. Since this is a differential system, these phase shifts are reference to the previous bit, not the absolute phase. Therefore, the previous bit is mapped to the reference phasor with zero degrees for every bit received, and the next bit is shown to have one of the 8 possible phase shifts referenced to 0 degrees. If this was not a differential system, then the phasors would be absolute, not referenced to the previous bit.

Thus, for the same bandwidth as BPSK, the D8PSK can send 3 times as many bits so that the bit rate is 3 times the bit rate as BPSK. The actual phase shifting occurs at the same rate as BPSK. The rate of the phase shifts is called the symbol rate and the actual bit rate in this case is 3 times the symbol rate. The symbol rate is important because it describes the spectral waveform of the signal in space. For example, if the symbol rate is 3 ksps, then the null-to-null bandwidth would be 6 kHz wide (2 times symbol rate) which would decode into 9 kbps.

2.7.1.9 Sidelobe Reduction Methods

One of the problems with phase shift keying systems is that the sidelobes can become fairly large and cause a problem with adjacent channel operation. The side lobes continue out theoretically to ± infinity. The main concern is usually the first or second sidelobes which are higher in magnitude. In order to confine the bandwidth for a particular waveform, a filter is required. The main problem with filtering a PSK signal is that this causes the waveform to be dispersed or spread out in time. This can cause distortion in the main signal and also cause more intersymbol interference (ISI) which is interference from adjacent pulses.

One type of filtering is called a raised cosine filter. This filter shapes the pulse with a raised cosine shape. This type of filter reduces the bandwidth of the signal but provides lower inter-symbol interference (ISI) than standard filters such as Butterworth or Chebychev.

Another type of filtering scheme using a Gaussian shaped pulse is called Gaussian MSK (GMSK). GMSK is a continuous phase modulation scheme that uses a Gaussian shaped pulse to do the MSK modulation or Gaussian shaped spectral filter. This modulation format reduces the sidelobe energy of the transmitted spectrum. The main

The Transmitter

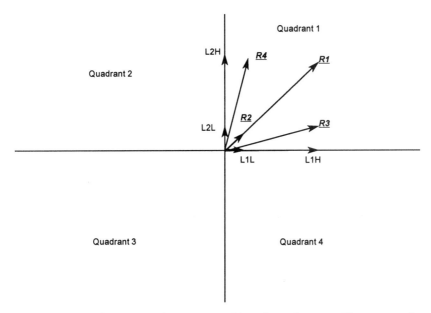

Figure 2-8 *One quadrant amplitude/phasor diagram for 16 OQAM.*

lobe is similar to MSK and is approximately 1.5 times wider than QPSK.

2.7.1.10 16 Offset Quaternary Amplitude Modulation (16OQAM)

Another modulation scheme is the 16 offset quaternary amplitude modulation (16OQAM) which is very similar to

OQPSK with both of the types of modulation identical except that each of the phasors have two amplitude states before summation. The resultant phasor has four different possibilities (R1, R2, R3, & R4) as shown in the phasor diagram Figure 2-8.

This shows only one of four possible quadrants available for simplicity. Since there are four possible amplitude/phase positions in one quadrant and a total of four quadrants, then there are 16 possible amplitude/phase positions in this modulation scheme.

Since this is 16 "offset" QAM, the "offset" means that the phasors in the defined quadrant can only change to the adjacent quadrants. For example, the phasors shown in quadrant 1 can only change to quadrants 2 and 4. These phasors cannot change from quadrant 1 to quadrant 3. In other words, only L1 or L2 can change at a time (see Figure 2-8).

2.7.1.11 Minimum Shift Keying (MSK)

If the OQPSK signal is smoothed by sinusoidally weighting the BPSK signals before summation, then another type of signal is produced called minimum-shift-keying (MSK). This process is shown in Figure 2-9.

The weighting frequency is equal to 1/2 the chip rate, so that it smoothes the transition of a +1,−1 by one cycle of the sinewave weighting signal. This provides a smoothing of the 180 degree phase shifts that occur in both of the channels. This smoothing slows the phase transitions of the two channels, which reduces the high frequency content of the spectrum that results in attenuation of the sidelobes that contain the high frequencies. Therefore, the spectrum is said to be efficient compared to standard PSK systems (BPSK, QPSK, etc.) since more power is contained in the main lobe and less in the sidelobes. The main lobe for MSK is 1.5 times larger than that for OQPSK for the same amount of bits sent since we started out with a spectrum twice as wide as the chip rate and taking the sum and difference of the modulating waveform and

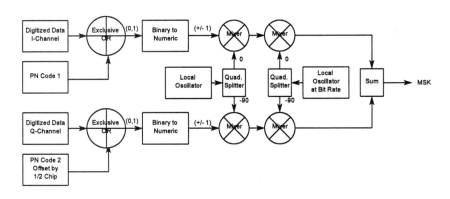

Figure 2-9 MSK generator.

putting them on top of each other. Since the modulation is 1/2 the chip rate, the spectrum is widened by a 1/2 chip width on both sides which equals 1 chip width bigger spectrum. Therefore, the total width would be 3 times the chip rate which is 1.5(2 × chip rate). MSK can be analyzed in two ways. The first is sinusoidally modulating the two BPSK channels in the OQPSK design before the summation takes place. This reduces the sidelobes and increases the main lobe. The other way to analyze MSK is using FSK and ensuring that the frequency separation is equal to the chip rate.

2.7.2 Frequency Shift Keying (FSK)

Another method of developing MSK is by frequency shift keying (FSK) and setting the frequency shift rate equal to

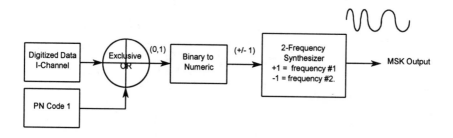

Figure 2-10 *MSK generator using FSK.*

The Transmitter

the frequency separation between the two frequencies. This is where minimum shift keying got its name since it is the minimum spacing between the two frequencies that can be accomplished and still recover the signal with a given shift rate. It is rather remarkable that the same waveform can be produced by two different methods of generation. Each of them provide a little more insight into MSK and two different ways to approach the understanding and design and analysis. A simple way of generating MSK is shown in Figure 2-10. A two frequency synthesizer is frequency shifted by the binary bit stream according to the PN code. If the rate is set to the minimum chipping rate, then MSK is the resultant output.

The frequency spacing of the FSK needs to be equal to the chip rate to generate MSK. The frequency shift provides one bit of information as does the BPSK waveform but the null-to-null bandwidth is only 1.5 times the chip rate as compared to 2 times the chip rate in the BPSK waveform. A simulation shows different types of FSK with different spacings with respect to the chip rate. If the frequency spacing is closer that the chip rate, then the information cannot be recovered. If the spacing is too far apart, the information can be retrieved using FSK demodulation techniques, but MSK is not generated as shown in Figure 2-11. As the frequency spacing approaches the chip rate, then the resultant spectrum is MSK, (see Figure 2-11).

2.7.3 Other Forms of Spread Spectrum Transmissions

Other forms of spread spectrum include frequency hopping, time hopping and chirped-FM. These all generate more bandwidth than is required to transmit the data over the link. The frequency hopping system generates the extra bandwidth by hopping over a wide frequency band. The time hopping system generates more bandwidth by the duty cycle of the time of transmissions. The chirped system generates more bandwidth by sweeping over a broad bandwidth. Then each system reduces the bandwidth in the detection process by reversing the process, thus providing process gain.

2.7.3.1 Frequency Hopped Transmitters

Most frequency hopped transmitters use a direct frequency hopped synthesizer for speed. Direct frequency synthesizers use a reference oscillator and directly multiply or translate the reference frequency to higher frequencies to be used in the hopping sequence. An indirect frequency synthesizer uses a phase lock loop (PLL) to generate the higher frequencies which slows the process due to the loop bandwidth of the PLL. A PN code is used to determine which of the frequency locations is to be hopped to. The frequency hop pattern is therefore pseudo-random to provide the spread spectrum and the processing gain required for the system. The process gain for a frequency hopped system is:

The Transmitter

G_p = # of frequency channels

However, the G_p assumes that the jammer does not jam multiple frequency cells. Usually the best jammer is a follower jammer. A follower jammer detects the frequency cell that the frequency hopper is currently in and immediately jams that frequency. The delay is associated with the time to detect the frequency and then to transmit the jammer on that frequency. For slow frequency hopping systems this methods works quite well because most of the time the signal will be jammed. Obviously, the faster the frequency hopping, the less effective the follower jammer has on the system. Also, many times a combination of frequency hopping and direct sequence is used, which slows down the detection process which reduces the effectiveness of the follower jammer.

2.7.3.2 Time Hopping

Time hopping entails transmitting the signal only at specified times. That is, the transmissions are periodic and the times to transmit are pseudo-random using a pseudo-random code. The process gain is equal to:

1/duty cycle

Basically, this means that if the duty cycle is short, on for a short period of time and off for a long period of time, the process gain is high.

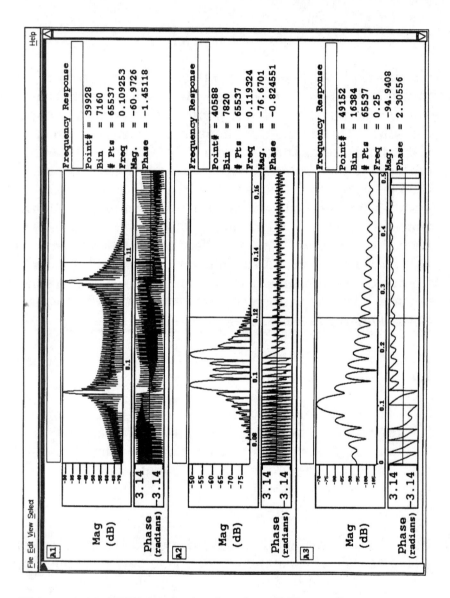

Figure 2-11 *FSK Analysis showing different frequency spacings.*

The Transmitter

2.7.3.3 Chirped-FM

Chirped-FM signals are generated by sliding the frequency in a continuously changing manner, similar to sweeping the frequency in one direction. The reason that these signals are called chirped signal is that when spectrally shifted to audible frequencies this signal sounds like a bird chirping. There are up chirps and down chirps depending on whether the frequencies are swept up in frequency or down in frequency respectively. The chirp signals can be generated by a sweeping generator with a control signal to reset the generator to the starting frequency at the end of the chirp. The chirp signals can also be generated by using Surface Acoustic Wave (SAW) devices excited by an impulse response (which theoretically contains all frequencies). The delay devices or fingers of the SAW device propagate each of the frequencies at a different rate thus producing a swept frequency output. This reduces the size of chirping hardware tremendously.

For chirp systems:

$$F = F_h - \Delta t \, (W/T)$$

where: W = Bandwidth of the dispersive delay line (DDL)
T = Dispersive delay time
F_h = highest freq in chirped bandwidth
Δt = delta time

For chirp systems the high frequencies have more attenuation for a given delay, so to end up with a flat (amplitude response) for all frequencies, the low frequencies need to be attenuated to match the high frequencies. Therefore, the total process has more insertion loss.

2.7.4 Multiple Users

To allow multiple users in the same geographical area, a scheme needs to be allocated to prevent interference between systems. Some of the schemes for reducing the interference are the following:

    a.   Time Division Multiple Access (TDMA)

    b.   Code Division Multiple Access (CDMA)

    c.   Frequency Division Multiple Access (FDMA)

These techniques are shown in Figure 2.12. Some control is required in all of these systems to ensure that each system in an operating area has different assignments.

TDMA provides interference reduction by having the systems communicate at different times or time slots at the same frequencies and codes (see Figure 2-12). If the systems have predetermined time slots, then the system is considered to be time division multiplexing. If a system accesses the time slot, say on a first come first serve basis, then the system is a true TDMA system.

The Transmitter

CDMA provides interference reduction by having the systems communicate on different codes, preferably orthogonal codes or Gold codes, at the same frequencies and time, which provides minimum cross correlation between the codes resulting in minimum interference between the systems (see Figure 2-12). A lot of study has been done to find the best code sets for this criteria.

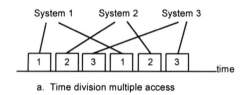
a. Time division multiple access

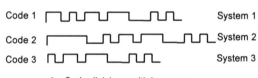
b. Code division multiple access.

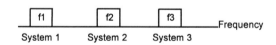
c. Frequency division multiple access.

Figure 2-12 *Multiple user techniques applying TDMA, CDMA, and FDMA.*

Generally, the shorter the codes, the more cross correlation interference is present, and fewer optimal codes can be obtained.

FDMA provides interference reduction by having the systems communicate on different frequencies and possibly at the same time and code (see Figure 2-12). This provides very good user separation since filters with very steep roll-offs can be used. Each user has a different frequency of operation and can communicate continuously on that frequency.

Each of the multiple user scenarios discussed above reduces interference and increases the communications capability in the same geographical area.

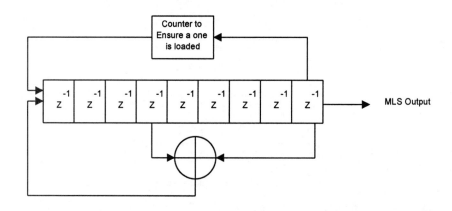

Figure 2-13 *Maximal length sequence generator.*

2.8 PN Code Generator

A method of building a PN generator consists of a shift register and a modulo-2 adder. There are several reference books that provide the correct taps to use in the modulo-2 adder to produce a maximal length sequence (MLS). If a MLS can be chosen so that only two taps are required, a simple two input, exclusive-or gate can be used. A "one" is loaded into the shift registers on power-up to get the code started (see Figure 2-13).

A counter is provided to ensure that the "all-zeros-case" is detected (which would stop the process) and a "one" is automatically loaded into the PN generator. The basic building block for generating direct sequence phase-shift keying systems is the maximal length sequence generator. There are other methods of generating a MLS codes. One process is to store the code in memory and then recall the code serially to generate the MLS code stream. This provides the flexibility to alter the code, for example make a perfectly balanced code of the same number of zeros as ones, which is a modification to the maximal length sequence generator. This provides a code with no DC offset. The DC offset in the system causes the carrier to be less suppressed. Also, other types of codes can be generated, for example the Gold codes that are used to provide orthogonal codes that reduce the cross-correlation between different code sets.

2.9 Summary

The transmitter is a key element in the design of the tranceiver. The transmitter provides a means of sending out the information, over the channel, with the power necessary to provide coverage to the intended receiver. The are many types of spread spectrum transmitters that provide process gain to reduce the effects of jammers and to allow more efficient use of the spectrum for multiple users.

2.10 References

[1] Jack K. Holmes, *Coherent Spread Spectrum Systems*, New York: Wiley & Sons, pp.251-267, 1982.

[2] Simon Haykin, *Communication Systems*, John Wiley & Sons Inc., New York, 1983.

[3] Scott R. Bullock, "Phase-Shift Keying Serves Direct-Sequence Applications", *Microwaves and RF*, December 1993.

Problems

1. Show that not filtering the output of a mixing product could result in a cancellation of the desired signal using carrier of 1 MHz and a signal carrier of .1 MHz.

2. If the unwanted sideband in problem 1 above was filtered, show that the desired signal can be retrieved.

3. Show by using a phasor diagram, that two QPSK modulators can be used to generate 8-PSK (8 phase states). Find the phase relationship of the two QPSK modulators?

4. What are the possible phase shifts using two OQPSK generators in problem 3 above?

5. Determine a 8-PSK modulator that eliminates the 180 degree phase shift. Why is the 180 degree shift a problem?

6. Why it is required to load a "1" in simple PN generator.

7. Find the process gain for a time hop signal with a duty cycle of 20%?

8. Find the process gain of a frequency hopper using 20 frequencies with a jammer eliminating two frequencies?

9. What is the basic difference between OQPSK and 16OQAM?

10. What is the basic difference between $\pi/4$DQPSK and D8PSK?

3

The Receiver

The receiver is responsible for downconverting, demodulating, decoding, and unformatting the data that is received over the link with the required sensitivity and bit error rate according to the link budget analysis of chapter 1. The receiver is responsible for providing the dynamic range to cover the expected range and power variations and to prevent saturation from larger power inputs and also provide the sensitivity for low level signals. The receiver provides detection and synchronization of the incoming

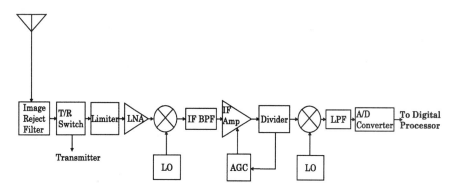

Figure 3-1 *Typical superheterodyne receiver.*

signals to retrieve the data that was sent by the transmitter. The receiver section is also responsible for despreading the signal when spread spectrum signals are used.

3.1 Superheterodyne Receiver

The main purpose of the receiver is to take the smallest input signal (minimum discernable signal (MDS)) at the input of the receiver and amplify that signal to the minimum discernable signal at the A/D converter while maintaining a maximum possible S/N ratio (SNR). A typical block diagram of the receiver is shown in Figure 3-1. Each of the blocks will be discussed in more detail. Most receivers are referred to as superheterodyne receivers which means that they use a common intermediate frequency (IF) and two stages to downconvert the signal to baseband as shown in Figure 3-1.

The reasons for using this type of receiver are:

a. A common IF can be used to reduce cost of parts and simplify design for different RF operating ranges.

b. Easy to filter image frequency since the image frequency is far away, by two times the IF frequency.

c. Easy to filter intermods and spurious responses.

The Receiver

d. Easy to design for tunable RF frequencies and frequency hopped signals using the synthesizer to change the frequencies, while the IF filter and circuitry remain the same.

Multiple stage downconverters using more than one IF frequency are sometimes used to aid in filtering and commonality. Occasionally a single downconverter is used which generally contains an image-reject mixer to reduce the effects of the image frequency.

3.2 Antenna

The receiver antenna gain is computed the same as the transmitter with the calculation performed using the link budget in Chapter 1. Factors to consider in determining the type of antenna to use are frequency, the amount of gain required, and the size. These factors are similar to the ones used to determine the transmitter antenna. Parabolic dishes are used frequently at microwave frequencies. In most systems, the receiver uses the same antenna as the transmitter which reduces cost and size of the system. The gain of the parabolic dish was calculated in the link budget section of Chapter 1. The antenna provides gain in the direction of the beam to reduce power requirements or to increase sensitivity of the receiver and also to reduce the amount of interference received. This is called selective jamming reduction. The gain of the antenna is usually expressed in dBi which is the gain in dB reference to what an isotropic radiator antenna would produce. This is the amount of amplifier gain that would have to be included in an isotropic radiator antenna to transmit the same amount of power in the same direction.

3.3 Transmit/Receive Control

As was described in the transmitter section, the signal is received by the antenna and passes through a device to allow the same antenna to be used by both the transmitter and receiver which can be a duplexer or diplexer, T/R switch, or circulator. The device chosen is required to provide the necessary isolation between the transmitter circuitry and the receiver circuitry and to prevent damage to the receiver during transmission.

3.4 Limiters

Limiters are placed in a receiver to protect the LNA from large spikes of energy. However, if the input power of the spike becomes too large, then the limiter no longer limits the input power of the spikes. Also, limiters take a finite time to respond which allows the spike of energy (ergs of power) to enter the receiver for a short period of time. Limiters also have an insertion loss that needs to be included in the losses in the link budget. This loss directly affects the link budget or the system noise figure.

3.5 Image Reject Filter

In many receivers an image reject filter is placed in the front end so that the image frequencies along with other unwanted signals are filtered out. These unwanted signals are capable of producing intermodulation products (intermods). The intermods are caused by the non-linearities of the LNA and the mixer. More discussion on intermods will be addressed later in this chapter.

The image frequencies are the other band of frequencies that when mixed with the LO will fall in the passband of the IF band. For example, if the desired input frequency is 900 MHz and the LO is at 800 MHz, the IF frequency is 100 MHz (900 – 800 MHz). The image frequency would be 700 MHz which produces the same IF frequency of 100 MHz (800 – 700 MHz). Therefore, if the receiver did not have an image reject filter in the front end, then an interference signal at 700 MHz would fall in the IF passband and jam the receiver.

Another approach to rejecting the image frequency is to use an image reject mixer. This is a specially designed mixer that is used to attenuate the image frequency.

3.6 Dynamic Range/Minimum Discernable Signal

There is a great deal of confusion in the definitions and measurements of minimum discernable signal (MDS) and dynamic range (DR) and so careful considerations must be done to ensure accurate analysis and to compare the systems with the same criteria. One of the main problems is that there are many systems today that process signals in digital signal processing (DSP) integrated circuits which makes comparison with the analog type systems difficult. The analog system generally does all of the processing up to the A/D, and then the output is detected and output digitally. The MDS is a measurement of how well a receiver can detect a signal with respect to noise and is generally calculated at the A/D converter. The DSP systems may not be able to evaluate the signal at the A/D, especially if spread spectrum systems are used and the process gain is performed in the digital domain. Before determining the MDS for a particular system, careful analysis needs to be performed to provide the optimal place in the system to do the calculation. Also, the criteria for calculating the MDS needs to be considered to evaluate each system architecture. Bit error rate, tangential sensitivity, S/N and others can be used as the criteria. However, to evaluate each system fairly, the same process should be used for comparison.

The minimum signal that the A/D converter can detect is calculated using the maximum voltage that the A/D can process. This assumes that there is no saturation and the A/D can handle the full dynamic range. For example, if the maximum voltage is 2 volts, then the maximum 2 volts will fill up the A/D converter with all ones. For each bit of the A/D converter, the dynamic range is increased by approximately 6 dB/bit. The reason for this is shown by the equation below:

The Receiver   97

$$10\log(1/2)^2 = -6 \text{ dB} \qquad 3.1$$

For each bit, the bit could be high or low which splits the decision in 1/2. For example, if there is a range from 0-10 and there is one bit with a threshold set at 5, then the range has been reduced by 1/2. If the bit is high, then the range is 5-10 and if the bit is low, then the range is from 0-5. Since voltage levels are being evaluated, the 1/2 is squared. If there are two bits, then the range can be divided into 4 levels, and each level is 1/4 the total range which calculates to −12dB.

Take for example, a system with an A/D that can handle 2 volts maximum and uses an 8-bit A/D converter. The 8-bit A/D converter provides 48 dB of dynamic range which produces an minimum signal that is approximately −48 dB below 2 volts. Therefore, the minimum signal in volts would be approximately 7.8 mV.

The system noise, which establishes the noise floor in the receiver, is comprised of thermal noise (kTBF), and source noise. The values of k,T,B, and F are usually converted to dB and summed together. Additional noise is added to the system due to the local oscillator's phase noise, LO bleedthrough, reflections due to impedance mismatch, etc. Therefore, the noise floor is determined by:

Noise Floor = Thermal noise + source noise

Approximately 3-6dB more noise is added to account for source noise in a typical receiver. The noise floor and the minimum signal at the A/D is used to calculate the S/N and the MDS is calculated using a given criteria for evaluation such as the tangential sensitivity.

The input noise floor for the system is used to determine the amount of gain that a receiver must provide to optimize the detection process. The gain required from the receiver for most analog type receivers is the amount to amplify this noise to the threshold of the A/D's least significant bit. Therefore, the gain is:

G_r (dB) = minimum signal level at the A/D (dB) − input noise floor level (dB)

The dynamic range of the receiver is the dynamic range of the A/D unless additional dynamic range devices are included such as log amplifiers or AGC to handle the higher level inputs. In order to provide the sensitivity and the dynamic range requirements these additional dynamic range devices are necessary.

The dynamic range of the receiver depends on the portion of the receiver that has the smallest difference between the noise floor and saturation. Dynamic range is another often misunderstood definition. The definition used here is the total range of signal level that can be processed through the receiver without saturation of any stage and within a set BER or defined MDS level. This assumes that the signal does not change faster than the AGC loops in the system. If the signal changes instantaneously or very fast compared to the AGC response, then another definition is required, instantaneous dynamic range (IDR). The IDR is

The Receiver

the difference between saturation and detectability, given that the signal level can change instantaneously. The IDR is usually the dynamic range of the A/D since the AGC does not respond instantaneously.

In order to determine the receivers dynamic range, a look at every stage in the receiver is required. The noise figure does not need to be recalculated unless there is a very large amount of attenuation between stages or the bandwidth becomes wider with a good deal of gain. Generally, once the noise figure has been established, a look at the saturation point in reference to the noise level at each component can be accomplished. An even distribution of gains and losses is generally the best. A dynamic range enhancer such as an AGC or a log amplifier can increase the receiver's dynamic range, however, they may not increase the IDR.

Note that the dynamic range can be reduced by an out-of-band large signal which causes compression and also can mix in noise. An analysis or testing of the jamming signals should be done to determine the degradation in dynamic range of the receiver.

3.7 Types of Dynamic Range

Dynamic range (DR) in receiver design can designate either amplitude or frequency. The two dynamic ranges are generally related and frequently only one dynamic range is used, but the actual DR is given for a particular reason.

3.7.1 Amplitude Dynamic Range

If the concern for a design requires the output to be within a given amplitude signal level, then the DR is generally given as the difference between the MDS to the maximum signal the receiver can handle (saturation).

Often times the 1 dB compression point provides a means of determining the maximum signal for the receiver. The 1 dB compression point is where the output signal power is 1 dB less that what the linear output power is expected to be with a given input. This means there is some saturation in the receiver (see Figure 3-2). Operating at this point results in 1 dB less power than what was expected.

If the dynamic range needs to be improved, then a log amplifier or an AGC amplifier can be added. The amplitude dynamic range is probably the most common used dynamic range term, and is easily understood by most people. Compression and saturation is important to prevent distortion of the incoming signal which generally increases the bit error rate.

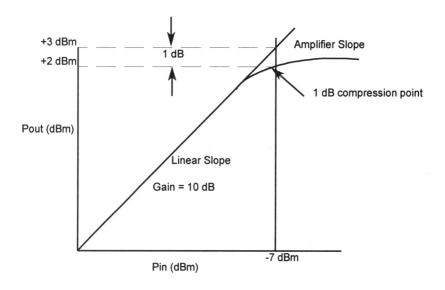

Figure 3-2 *1 dB compression point.*

3.7.2 Frequency Dynamic Range

Many systems are concerned with the frequency dynamic range. This is the ability of a receiver to distinguish different frequencies whether it be between multiple desired frequencies or between the desired frequency and the spurious signals or jammers. This is related to the 1 dB compression point DR since when non-linearities are present, there are unwanted spurious signals generated which reduce the frequency DR (see Figure 3-3).

There are two methods of analysis that generally prove useful when considering frequency DR.

3.7.2.1 Single Tone Frequency Dynamic Range

One method is to use a one tone DR and calculate the spurs that will reduce the frequency DR because of this tone. This method is not used often because generally there is more than one input signal.

3.7.2.2 Two Tone Frequency Dynamic Range

A better way, and one that is more commonly used, is to define DR using a two tone DR analysis. This approach takes two input signals and their generated spurious products and calculates their power levels using intercept points on the linear or desired slope (see Figure 3-3).

The second order intercept point helps to determine the second harmonics of the two inputs (2×0, 0×2) and product of the two signals (1×1). The second order spurs are the highest power intermod products that are created due to a non-linear portion of the receiver. However, if the bandwidth is limited to less than an octave, these products are eliminated and the third order spurs (2×1, 1×2) become the strongest products in band. These spurs generally fall

The Receiver

into the operating band, especially when they are close together in frequency.

The third order intercept point becomes the criteria for calculating the DR for the receiver (see Figure 3-3). Caution must be taken when performing the analysis

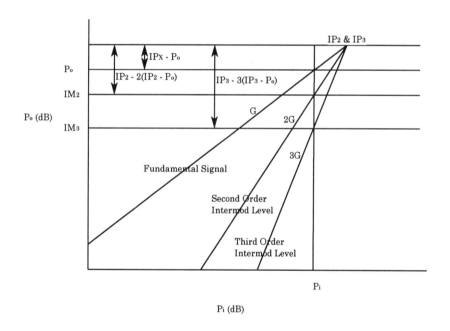

Figure 3-3 *Graphical analysis of IM equations.*

because if the bandwidth is greater than an octave, the second order intermods cannot be neglected.

3.8   Second and Third Order Intermodulation Products

The second and third order intermod points are used to determine the intermod levels in the system. The output power in dB plotted against the input power is a linear function up to the compression point. The slope of this line is constant with gain, that is P_o (dBm) = P_i (dBm) + G(dB). The slope of the second order line is equal to twice that of the fundamental. That is, for every 1 dB of increase of output fundamental power, the second order output signal power increases by 2 dB. The slope of the third order line is equal to three times that of the fundamental. That is, for every 1 dB of increase of output fundamental power, the third order output signal power increases by 3 dB. If the fundamental plot was extended linearly beyond the 1 dB compression, then this line would intersect the second and third order lines. The point where the second order curve crosses the fundamental is called the second order intercept point. The point where the third order curve crosses the fundamental is called the third order intercept point. The intercept points are given in dBm. The intermod levels can then be calculated for a given signal level and can be graphically shown and calculated as shown in Figure 3-3.

The second order intermod signal will be 2 times farther down the power scale from the intercept point as the fundamental. Therefore, the actual power level in dB of the second order intermod would be:

$$IM_2 = IP_2 - 2(IP_2 - P_o) \qquad 3.2$$

where:

IM_2 = the second order intermod power level in dBm
IP_2 = second order intercept point
P_o = output power of the fundamental signal

The third order intermod signal will be 3 times farther down the power scale from the intercept point as the fundamental. Therefore, the actual power level in dB of the third order intermod would be:

$$IM_3 = IP_3 - 3(IP_3 - P_o) \qquad 3.3$$

where:

IM_3 = the third order intermod power level in dBm
IP_3 = third order intercept point
P_o = output power of the fundamental signal

For example, suppose the second order and third order intercept points were at +30 dBm, the input power level is at −80 dBm, and the gain is at +50 dB. The signal output level would be at −30 dBm. The difference between the intercept points and the signal level is 60 dB. The second order intermod signal level would be:

$$IM_2 = +30 \text{ dBm} - 2(60 \text{ dB}) = -90 \text{ dBm} \qquad 3.4$$

The third order intermod signal level would be:

$$IM_3 = +30 \text{ dBm} - 3(60 \text{ dB}) = -150 \text{ dBm} \qquad 3.5$$

Since the second order signals can be generally filtered out, the third order products are the ones that limit the dynamic range of the system.

3.9 Calculating Two Tone Frequency Dynamic Range

A way of calculating DR is shown in Figure 3-4. This example is for third order frequency DR. This method compares the input power levels for calculating DR. The input power that causes the output power P_o to be right at the noise floor is equal to P_1. This is the minimum detected input signal that can theoretically be detected. The input power, P_3, is the power that produces third order intermod products at the same output power P_o which is at the noise floor.

An increase in input power will cause the third order spurs to increase above the noise (note that the spurs will increase in power faster than the desired signal level as input power is increased).

The equations below are simple slope equations:

$$(P_o - IP_3)/[P_1 - (IP_3 - G)] = 1$$

The Receiver

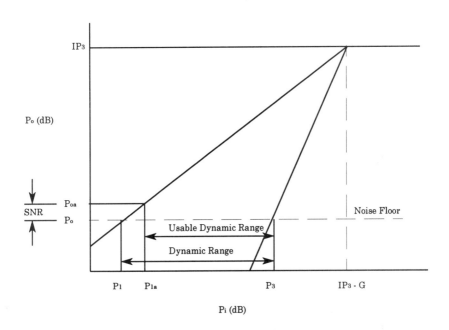

Figure 3-4 *Frequency dynamic range analysis.*

$$P_1 = P_o - G$$
$$(P_o - IP_3)/[P_3 - (IP_3 - G)] = 3$$
$$P_3 = (P_o + 2IP_3 - 3G)/3 \qquad 3.6$$

The difference in the input powers gives the DR of the system:

$$DR = P_3 - P_1 = (P_o/3 + 2IP_3/3 - G) - (P_o - G) = 2/3(IP_3 - P_o) \quad 3.7$$

$$P_o = \text{noise floor} = -174 + 10\log BW + NF + G - PG \quad 3.8$$

$$DR = 2/3(IP3 + 174 - 10\log BW - NF - G + PG) \quad 3.9$$

where:

> DR = dynamic range
> IP_3 = third order intercept point
> BW = bandwidth
> NF = noise figure
> G = gain
> PG = process gain

Before the detector, the receiver noise figure bandwidth uses the smallest IF bandwidth for the calculation.

The above analysis gives the theoretical DR, but for usable DR a SNR needs to be specified. This requires the input power for the desired signal to increase sufficiently to produce this desired SNR (see Figure 3-4). Since the desired spur level is still at the noise floor, this is a direct subtraction (dB) of the desired SNR as shown:

$$DR_a = P_3 - P_{1a}$$
$$= 2/3(IP_3 + 174 - 10\log BW - NF - G + PG) - SNR \quad 3.10$$

From the same analysis, the second order usable DR is:

$$DR_a = 1/2(IP_2 + 174 - 10\log BW - NF - G + PG) - SNR \quad 3.11$$

The Receiver   109

This is the third order spurious-free dynamic range (SFDR).

3.10 System Dynamic Range

A main concern in a receiver system analysis is the determination of where the minimum dynamic range occurs. In order to easily determine where this is, a block diagram of the receiver with the noise level and the saturation level is developed. This can be included as a parameter in the link budget, however, instead of cluttering the link budget a separate analysis is usually performed. A list of the outputs of the devices are listed with the noise level, the saturation level and the calculated amplitude dynamic range. This is shown in the spread sheet, (Table 3-1).

The procedure is to list all of the power levels and noise levels at every point in the system. A good point to start is at the LNA where the noise figure is usually established. This establishes the relative noise floor. Note that this is using the bandwidth of the devices so far. If the bandwidth is narrowed down the line then the noise is lowered by 10logBW. The top level is calculated by either using the 1 dB compression point or the third order intercept point.

Table 3-1 *Receiver dynamic range calculations.*

RECEIVER ANALYSIS FOR OPTIMUM DYNAMIC RANGE					
Receiver	Gain/Losses	Sat.(dBm)	Noise(dB	IDR (dB	DR (dB)
RF BW (KHz)(KTB)	100.00		-124		
RF Components	2.00	50.00	-124	174.00	174.00
LNA (In)		7.00	-124	131.00	131.00
LNA Noise Figure (int.)	3.00		-121.00		
LNA (Out)	30	7	-91.00	98.00	98.00
Filter (In)		50	-91.00	141.00	141.00
Filter (Out)	-2	50	-93.00	143.00	143.00
Mixer (In)		0	-93.00	93.00	93.00
Mixer (Out)	-10	0	-103.00	103.00	103.00
IF Amp(In)		7	-103.00	110.00	110.00
IF Amp (Out)	50	7	-53.00	60.00	60.00
IF Filter (In)		50	-53.00	103.00	103.00
IF Bandwidth (kHz)	10		-63.00		
IF Filter (In)		50	-63.00	113.00	113.00
IF Filter Loss	-2	50	-65.00	115.00	115.00
Baseband Mixer (In)		0	-65.00	65.00	65.00
Baseband Mixer (Out)	-10	0	-75.00	75.00	75.00
A/D Converter(bits)	8	-27	-75.00	48.00	48.00

For the 1 dB compression point, the DR is calculated by taking the difference of the relative noise floor and this point. For the third order intercept point method, the DR is calculated by taking 2/3 of the distance between the relative noise floor and the intercept point. The minimum distance is found and from this point the gains are added (or subtracted) to determine the upper level input to the system. The difference between the input relative noise and the upper level input is the dynamic range.

The Receiver

Table 3-2 Receiver dynamic range calculations using feedback AGC.

RECEIVER ANALYSIS FOR OPTIMUM DYNAMIC RANGE WITH AGC					
Receiver	Gain/Losses	Sat.(dBm)	Noise(dB	IDR (dB	DR (dB
RF BW (KHz)(KTB)	100.00		-124		
RF Components	2.00	50.00	-124	174.00	174.00
LNA (In)		7.00	-124	131.00	131.00
LNA Noise Figure (int.)	3.00		-121.00		
LNA (Out)	30	7	-91.00	98.00	98.00
Filter (In)		50	-91.00	141.00	141.00
Filter (Out)	-2	50	-93.00	143.00	143.00
Mixer (In)		0	-93.00	93.00	93.00
Mixer (Out)	-10	0	-103.00	103.00	103.00
IF Amp(In)		7	-103.00	110.00	110.00
AGC	30				
IF Amp (Out)	50	7	-53.00	60.00	90.00
IF Filter (In)		50	-53.00	103.00	133.00
IF Bandwidth (kHz)	10		-63.00		
IF Filter (In)		50	-63.00	113.00	143.00
IF Filter Loss	-2	50	-65.00	115.00	145.00
Baseband Mixer (In)		0	-65.00	65.00	95.00
Baseband Mixer (Out)	-10	0	-75.00	75.00	105.00
A/D Converter(bits)	8	-27	-75.00	48.00	78.00

The overall input third-order intercept point is found by adding 1/2 the input DR (refer back to the 2/3 rule where 1/2 of 2/3 is 1/3 so the upper level input is 2/3 of the intercept point).

Adding an AGC extends the DR of the receiver generally by extending the video and detection circuitry as shown in Table 3-2.

In order to determine the required AGC, two main factors need to be considered:

 a. Dynamic range of the A/D converter.

 b. Input variations of the received signals.

The dynamic range of the A/D converter is dependent on the number of bits in the A/D. For an 8-bit A/D converter the usable DR with a SNR = 15 dB is approximately 30 dB (theoretical is approximately 6dB/bit = 48 dB). This establishes the range of signal levels that can be processed effectively. If the input variations of the received signals is greater than this range, then an AGC is required. If the instantaneous dynamic range (IDR) needs to be greater than the A/D range, then a feed-forward AGC needs to be implemented to increase IDR. The amount of feed-forward AGC required is calculated by subtracting the A/D range from the input signal range.

The feedback AGC extends the dynamic range but does nothing to the instantaneous DR as shown in Table 3-2. The IDR tests the range in which the receiver can handle an instantaneous change (much quicker than the AGC response time) in amplitude on the input without saturation or loss of detection.

The Receiver 113

The IDR can be extended by using a feed-forward AGC or a log amplifier. The problem with a log amp is that the IDR is extended as far as amplitude but since it is non-linear the frequency IDR may not improve, and might even worsen. The feed-forward AGC is faster than the signal that is traveling through the receiver, and therefore, the IDR is improved for both amplitude and frequency and can be done linearly.

3.11 Tangential Sensitivity

For pulse systems, since the MDS is hard to measure, another parameter is measured called the tangential sensitivity (TSS) and is shown in Figure 3-5.

The signal power is increased until the bottom of the pulse noise is equal to the top of the noise level in the absence of a pulse. A tangential line is drawn to show that the pulse noise is tangential to the rest of the noise. With this type of measurement, at least a comparison between receivers can be established. The tangential sensitivity measurement is good for approximately ± 1dB.

Dynamic range is critical in analyzing receivers of all types. Depending on the particular design criteria, both amplitude and frequency dynamic range are important in the analysis of a receiver. Methods of determining and calculating dynamic range need to be incorporated in the design and analysis of all types of receivers. This is essential in optimizing the design and calculating the minimizing factor or device that is limiting the overall receiver dynamic range. Graphical methods for calculating

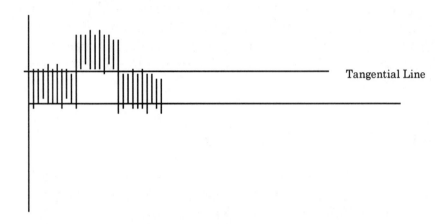

Figure 3-5 *Tangential sensitivity (TSS) for a pulse system.*

dynamic range provide the engineer with a useful tool in analyzing receiver design. Both the dynamic range and the instantaneous dynamic range need to be considered for optimal performance. AGC can improve the receivers dynamic range tremendously but generally does not affect instantaneous dynamic range. Feedforward AGC and a log amp improve the instantaneous dynamic range (IDR), however, the log amp is non-linear and only improves the amplitude IDR and actually degrades the frequency IDR. The minimum discernable signal (MDS) is hard to define and measure. Generally a SNR is specified to help define the MDS. For pulse systems, the tangential sensitivity is utilized to improve the measurement of MDS and to compare receivers on a fairly accurate criteria.

3.12 Low Noise Amplifier

A low noise amplifier (LNA) is used in a receiver to establish the noise figure (NF) providing that the losses between gain elements are not too large and the bandwidth does not increase. The NF is calculated as follows:

Noise Factor $F_s = F_1 + (F_2-1)/G_1 + (F_3-1)/G_1G_2$ .. 3.12
note: F is in actual gain and not in dB.

$NF = 10\log F_s \approx 10\log F_1$, if G_1 is large 3.13

For example, given the following parameters:

NF_1 = 6 dB(LNA), L = 8 dB(mixer, insertion loss), NF_2 = 6 dB, for the IF amplifier with 20 dB of gain.

G_1 = 25 dB(LNA) including losses
F_1 = 4, F_2 = 4, L = 6.3
F_s = 4 + 6.3–1/(316/6.3) = 4.1, NF_s = 6.1 dB

Without an LNA, assuming no contribution after the second amplifier:

$F_s = L \times F_2 = 6.3 \times 4 = 25.2$, NF_s = 14 dB

The mixer noise figure (conversion loss) in dB can be added to the next stage amplifier's noise figure in dB or you can solve for the noise factor as above for no LNA as shown:

$$F_s = F_1 + (F_2-1)L_1 + (F_3-1)L_1/G_2 \qquad 3.14$$

Note, however, that G_1 is now the conversion loss L_1 which is the reciprocal of G_1. Since F_1 is equal to L_1 and using only the first two terms, the equation becomes:

$$F_s = F_1 + (F_2-1)F_1 = F_1(1 + F_2 - 1) = F_1 F_2 \qquad 3.15$$

that is the two noise factors multiplied together as shown above, which is the same as adding the noise figures in dB.

The mixer can be treated as a loss between the antenna and the first amplifier that establishes the noise figure for the system. The mixer does not really add noise to the system to raise the noise figure, but attenuates the signal with respect to the noise floor (set by the temperature or kTB) which reduces the SNR and thereby changes the system noise figure.

In general, the VSWR for a mixer is not very good, about 2:1. If the source is 2:1, then, for electrical separation of greater than 1/4 wavelength, the VSWR can be as much as 4:1. This is equivalent to having a mismatched load of 4 times larger (or 1/4 as large) than the nominal impedance.

The LNA also provides isolation between the LO used for down-conversion and the antenna. This prevents LO

The Receiver   117

bleedthrough from appearing at the antenna port. However, a duplexer generally solves this problem.

Another benefit for using an LNA in the receiver is that less power is required from the transmitter if the noise figure is low. For a given desired S/N or E_b/N_o, the noise figure alters these according to the link budget as specified in Chapter 1.

In general, the LNA amplifies the desired signal with minimum added noise and establishes the receiver noise level.

Some reasons for not using the LNA in a system are as follows:

> (1) Preamplifiers reduce dynamic range of a receiver. For example, if the 1 dB compression point is +10 dBm, the maximum input power to the LNA with a gain of 25 dB is −15 dBm. Without a preamp, the 1 dB compression for a low level mixer is +3 dBm. Also, the LO drive can be increased to provide even a higher dynamic range without the LNA.

> (2) The interference signals, clutter, and bleedthrough from the transmitter, etc., may be more significant than the noise floor level and therefore, establishing a lower noise figure with the LNA may not be required.

(3) Cost, space, and weight are less without a preamplifier.

The results show that the LNA gives a better noise figure which results in better sensitivity at the cost of reduced dynamic range, cost and space. The LNA also provides isolation between LO and the antenna. Also, the low NF established by the LNA decreases the power requirement for the transmitter. However, some airborne radar systems use mixer front ends (no preamp) to reduce weight.

3.13 Downconversion

After the LNA, most receivers need to be downconverted to a lower frequency range for processing. This is accomplished by using a local oscillator (LO), a mixer, and a band pass filter (BPF) or low pass filter (LPF) to select the difference frequency as shown in Figure 3-1.

Some systems get converted to baseband in the process and others include a intermediate frequency (IF) which requires a double downconversion. Double downconversion requires an additional LO, mixer, and filter to convert the signal to baseband. The double downconversion relaxes constraints on filters and allows for common circuitry for different systems. For example, a receiver operating at 1 GHz and another operating at 2 GHz could both down convert the signal to a common IF frequency of 50 MHz and the rest of the receiver could be identical.

The Receiver

Some of the newer receivers digitize the IF band directly and do a digital baseband downconversion before processing. This is especially true where quadrature downconversion (I&Q) is required. This relaxes the constraints of quadrature balance between the quadrature channels in the IF/analog portion of the receiver since it is easier to maintain quadrature balance in the digital portion of the receiver, and it reduces hardware and cost.

3.14 Splitting Signals into Multiple Bands for Processing

Some receivers split the incoming signal into multiple bands to aid in processing. This is common with intercept type receivers that are trying to determine what an unknown signal is. If the signal is divided up in multiple bands for processing, the signal is split up but the noise remains the same if the bandwidth has not changed. Therefore, the S/N will degrade each time the signal is split by at least 3dB. However, if the signal is coherent, and is summed coherently after the split, then there is no change in the S/N except some loss in the process gain.

Therefore, careful examination of the required S/N and the type of processing used with this type of receiver will aid in the design.

3.15 Phase Noise

Phase noise is the noise generally associated with oscillators or phase locked loops. Phase noise is the random change in the phase or frequency of the desired frequency or phase of operation. The noise sources include thermal noise, shot noise, and flicker noise and 1/f noise. Halford discovered that the 1/f noise could be suppressed by using a 10 to 300 ohm unbypassed emitter resistor. This noise is less than the corner frequency and as a rule of thumb is −120 dBc/Hz at 1 Hz offset [10].

The different types of phase and frequency noise and the causes that generate each of the processes and definitions are shown in Figure 3-6. The noise beyond about 10 kHz is dominated by white noise which is generally referred to as kTB noise since it is derived using Boltzman's constant, temperature, and bandwidth. The 1/f noise or flicker noise is the dominant noise source from about 100 Hz to 10 kHz. This noise is caused by noisy electrical parts such as transistors and amplifiers. The next type of noise that is the dominant noise from about 20 Hz to 100 Hz is passive resonator devices such as cesium and rubidium standards. This is known as white FM noise since it appears to change the frequency with respect to white noise. The next type of noise is the physical resonance mechanism or the actual parts in oscillators and is close to the carrier, about 5 Hz to 20 Hz. This type of noise is known as flicker FM. The closest noise to the actual carrier frequency is the 1/ffff noise and is caused by vibration, shock, temperature, and other environmental parameters. This is called random walk FM and is very difficult to measure since it is so close to the carrier and requires very fine resolution.

3.16 Mixers

Since the downconversion process requires a mixer, there are questions to be asked to ensure the right mixer is selected. Some of the questions are:

    a.   High or low level mixers?

    b.   High side or low side injection?

    c.   Isolation--LO bleedthrough?

    d.   Level of spurious responses and intermods?

High level mixers use higher voltages and require more power to operate than do the standard low level mixers. The advantage of using high level mixers are:

    a.   High 3rd order intercept point [less cross-modulation, more suppression of 2-tone 3rd order response, $(2f_1 - f_2) \pm LO$].

    b.   Larger dynamic range.

    c.   Lower conversion loss, better noise figure.

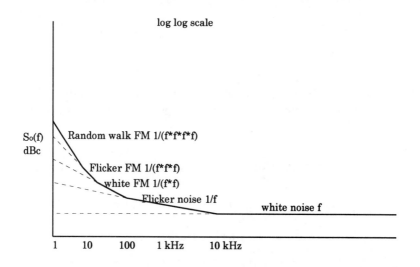

Figure 3.3 Phase Noise

$1/(f*f*f*f)$ = close to carrier difficult to measure, vibration, shock, temperature, environmental.

$1/(f*f*f)$ = observable in high quality oscillators, masked in low quality oscillators, not fully understood, physical resonance mechanism or actual parts in the oscillator.

$1/(f*f)$ = common in passive-resonator like cesium and rubidium standards.

$1/(f)$ = transistors, amplifiers, etc., noisy electronics, LNA helps.

f = probably produced like the 1/f noise, stages of amplification is mainly responsible, broadband noise.

Notes: 10log(script L) = dBc/Hz
deviation = (delta f)/(f rate)

Figure 3-6 *Phase noise analysis.*

d.   Best suppression above top two rows of a

mixer spurious chart.

The standard low level mixers are used in most systems. The advantages of the low level mixers are the following:

- a. Less complex, easier to balance giving better isolation, lower DC offset and less mixer-induced phase shift.

- b. More covert, less LO bleedthrough.

- c. Best suppression of bottom two rows of mixer chart.

- d. Less system power and less expensive.

Depending on the receiver application, these advantages need to be reviewed in order to make the best selection of mixers.

Another consideration when selecting a mixer is whether to use high side vs. low side injection. High side refers to the LO being at a higher frequency than the input signal frequency and low side refers to the LO being at a lower frequency than the input signal frequency. Some of the main points to consider in the selection are:

- a. Performing a spurious signal analysis assists in determining the best method to use.

b. High side injection inverts the signal; lower sideband becomes upper sideband which may cause some problems.

c. Low side provides a lower frequency for the LO which may be easier to get and less expensive.

d. Image frequency is different and analysis needs to be done to determine which of the image frequencies will affect the receiver the most.

Image reject mixers are sometimes used to reduce the image frequency without filtering. This is useful in a receiver system that contains a wide bandwidth on the RF front end and a narrow bandwidth IF after the mixer so that the image frequency cannot be filtered. This eliminates the need for double downconversion in some types of receivers. However, double downconversion rejects image frequency much better. This is generally a tradeoff on the image frequency rejection and cost.

3.16.1 Mixer Spur Analysis

A mixer spur analysis is done for each mixer output to:

a. Determine which mixer products fall in the passband of the output.

The Receiver   125

    b.   Determine the frequencies to be used in order to ensure than mixer spurs do not fall in the pass band.

Spurious signals are generated when mixing signals up and down in frequency. They are the mixer products and are designated as n × m order spurs where n is the harmonics of the LO and m is the harmonics of the RF or IF (the input of the mixer could be an RF signal or an IF signal depending on the frequency translation). These mixer spurs can cause problems if they fall in the passband since they cannot be filtered. A spurious analysis should be done to determine where the spurs are located, if they fall in the passband, and the power of the spur with respect to the desired signal. Several software application programs have been written to assist in determining where the spurs are for a given system. These programs take two input frequencies and multiply them together depending on the order specified to generate the possibilities. For example, given this third order system:

        1×0 =   0×1 =
        2×0 =   0×2 =
        3×0 =   0×3 =
        1×1 =   1×1 =
        2×1 =   1×2 =

The first number multiplies the LO and the second number multiplies the input and then the resultants are added and subtracted to determine the frequency of the spurs. Note that the 1×1 contains the desired signal and a spur depending whether or not the wanted signal is the sum or the difference. Usually this is obvious and the unwanted signal is filtered out.

Selecting the right LO or specifying the operational frequencies for a selected level of spur rejection is done by comparing all the possible spur locations with respect to a given amplitude threshold. This is dependent on highside,

lowside, sum or difference. Frequency selection depends on mixer spur analysis.

3.16.2 Six Order Analysis

A six order spur analysis is standard for spur analysis. This ensures that the spurious signals are at least 60 dB down from desired output for most mixers. The analysis below is for a six order spurs analysis with the desired signal being the difference frequency. The LO is higher in frequency that the highest desired frequency, which eliminates the spurs, 21, 31, 41, 51, 32, and 42.

<u>SPUR</u>

10    $LO > LO - f_l$ always and for 20,30,40,50,60.   3.16
       $LO < LO - f_h$ never and for 20,30,40,50,60.

01    $f_l > LO - f_l$ $LO < 2f_l$ 02,03,04,05,06 less restricted.

12    $2f_l - LO > LO - f_l$ Therefore: $LO < 3/2f_l$, $LO > f_h$: Spurs 13,14,15 are less restricted.

The Receiver

$2f_h - LO < LO - f_h$ Therefore: $LO > 3/2f_h$:15 gives $LO > 3f_h$, assumed $2f_h > LO$ impossible.
$LO - 2f_l > LO - f_l$ Therefore: Impossible.
$LO - 2f_l < LO - f_h$ Therefore: $LO > 2f_l$, $f_l > 1/2f_h$
Spurs 13,14,15 less restricted.

22   Since $LO > f_h$ then there are only two possibilities.
$2LO - 2f_h > LO - f_l$ Therefore: $LO > 2f_h - f_l$
$2LO - 2f_l < LO - f_h$ Therefore: $LO < 2f_l - f_h$

23   $3f_l - 2LO > LO - f_l$ Therefore: $LO < 4/3f_l$
$3f_h - 2LO < LO - f_h$ Therefore: $LO > 4/3f_h$, $LO < 3/2f_h$

$2LO - 3f_l < LO - f_h$ Therefore: $LO < 3f_l - f_h$, $LO > 3/2f_l$
$2LO - 3f_h > LO - f_l$ Therefore: $LO > 3f_h - f_l$, $LO > 3/2f_h$

33   $3LO - 3f_h > LO - f_l$ Therefore: $LO > (3f_h - f_l)/2$
$3LO - 3f_l < LO - f_h$ Therefore: $LO < (3f_l - f_h)/2$

24   $4f_l - 2LO > LO - f_l$ Therefore: $LO < 5/3f_l$
$4f_h - 2LO < LO - f_h$ Therefore: $LO > 5/3f_h$, $LO < 2f_h$
$2LO - 4f_l < LO - f_h$ Therefore: $LO < 4f_l - f_h$, $LO > 2f_l$
$2LO - 4f_h > LO - f_l$ Therefore: $LO > 4f_h - f_l$, $LO > 2f_h$

06   $6f_h < LO - f_h$ Therefore: $LO > 7f_h$
01,02,03,04,05 less restricted.

Results are shown below:

1. $LO > 7f_h[06]$ and $f_l > f_h/2[12]$

2. $f_h < LO < 2f_l - f_h$ impossible

3. Several possibilities exists and need to be analyzed.

The general solution is (1) above. This requires the LO to be greater than 7 times the highest input frequency (required to eliminate the worst case 06 spur) and the highest input frequency is less than twice the lowest input frequency (required to eliminate the worst case 12 spur). The second equation is impossible and the third solution may create specific bands of operation. The same type of mixer analysis can be done for each mixer configuration to ensure that the spurious signals do not fall in the desired band and can be filtered out from the system.

3.17 Bandwidth Constraints

The bandwidth of the transmitted and received signal can influence the choice of IF bands used. The IF bandwidth needs to be large enough so that there is no signal foldover and to prevent any aliasing. Also, the bandwidths need to be selected to relax design constraints for the filters.

For example, a direct sequence spread bandwidth is approximately 200 MHz wide. Since the total bandwidth is 200 MHz, the RF center frequency needs to be greater than twice the 200 MHz bandwidth so that the unwanted sidebands can be filtered. An RF of 600 MHz is chosen because of availability of parts. This also provides sufficient margin for the filter design constraints. If the bandwidth is chosen so that the design constraints on the filters require the shape factor to be too large, then the filters may not be feasible to build.

Bandwidth plays an important function in the S/N ratio of a system. The narrower the bandwidth, the lower the noise floor since the noise floor of a receiver is related to kTBF where B is the bandwidth.

3.18 Filter Constraints

The shape factor is used to specify a filter's roll-off characteristics. The shape factor SF is defined as:

$$SF = f(-60dB)/f(-3dB) \qquad 3.17$$

> Note: These values may be evaluated at different levels for different vendors, check with the company.

Typical achievable shape factors range from 2 to 12, although larger and smaller shape factors can be achieved. Shape factors as low as 1.05 (60/3) can be achieved (crystal filters). Note that insertion loss generally increases as shape factor decreases. Insertion loss ranges from 1.2 to 15 dB. SAW filters achieve even a better shape factor at the expense of greater insertion loss and cost.

The percent bandwidth is another way to specify the filter's roll-off characteristics. The percent bandwidth is the percent of the carrier frequency. For example, a 1% bandwidth for 100 MHz would be 1 MHz bandwidth. This is usually specified as the 3 dB bandwidth. If the bandwidth is not specified, then the 3 dB bandwidth is used. Typical achievable bandwidths using crystal filters for 10 kHz-200 MHz range from .0001%-3%.

3.19 Pre-Aliasing Filter

A filter is required before any sampling function to prevent aliasing of the higher frequencies back into the desired signal bandwidth (see Figure 3-7). This filter is called a pre-aliasing filter. To illustrate the concept, if we had a desired signal at 1 MHz that we were sampling at a 2 Msps rate, the output of the sampler would be alternating ± values at a 2 Msps rate. The reconstruction of the samples would produce a 1 MHz frequency. If there is an incoming signal at a higher frequency and it was sampled at the same rate (undersampled), then the output would look like the same frequency when reconstructed which would interfere with the correct signal. For example, if the high frequency is unwanted and the sampling is at the Nyquist rate for the lower frequency, then this sampling of the higher frequency produces a lower frequency that falls into the lower bandwidth and interferes with the desired signal (see Figure 3-7).

The Receiver

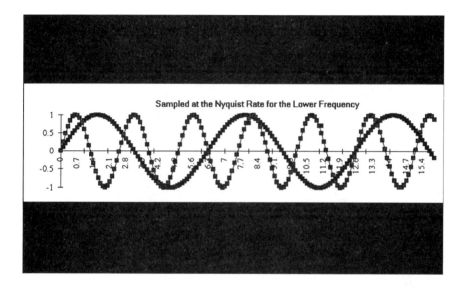

Figure 3-7 *Graph showing the Nyquist Criteria for Sampling.*

3.20 A/D Converter

The A/D converter is a device that samples the analog signal and converts it to a digital signal for digital processing. The A/D has to sample at least as fast as the Nyquist rate, or twice as fast as the highest frequency component of the signal. One of the problems that occurs when an analog signal is digitized is the resolution of the step size. The error associated with the step is called quantization error.

As the signal becomes smaller in an A/D converter, the error becomes larger as shown on Figures 3-8 and 3-9. Note that if the signal is large, the quantization errors in both the I and Q channels produce a small change in amplitude and phase. The phase error is just a few degrees and the amplitude is a very small percentage ofthe actual amplitude as shown in Figure 3-8. As for the smaller signal, quantization error in the I and Q channels produces a large error in phase, up to 45 degrees, and the amplitude error is from 0 to 1.414 as shown in Figure 3-9.

Most A/D converters are linear, which makes it harder to detect smaller signals. To improve the sensitivity of the A/D, a log amp is used before the A/D or an A/D with a log scale (µ-law). This gives a higher response (gain) to the lower level signal and a lower response (gain) to the high level signals.

The Receiver 133

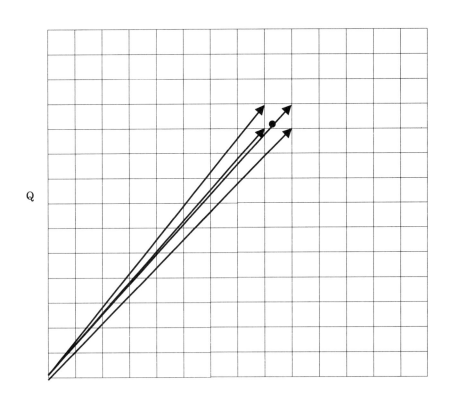

Figure 3-8 *A/D quantization error for large signals.*

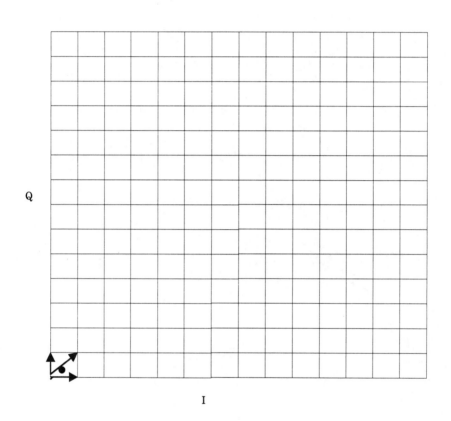

Figure 3-9 A/D quantization error for small signals.

The Receiver

Another way to improve the sensitivity of the lower level signals is to use a piecewise linear solution where there are multiple A/D stages with different scaling factors. Each finer resolution stage is set to cover the LSBs of the previous A/D. Figure 3-10 shows the results of this type of solution. This method provides a large dynamic range while still being a linear process. Also, a piecewise µ-law solution can be used to increase the dynamic range of the A/D, however, this would not be a linear solution.

3.21 Digital Signal Processing

Once the analog signal is converted by the A/D to a digital signal, the digital signal processor finishes the reception by interpreting the data.

Many receivers are using digital signal processing (DSP) technology to accomplish most of the detection function. The RF/analog portion of the receiver is only used to down-convert the signal to a lower IF or baseband, spectrally

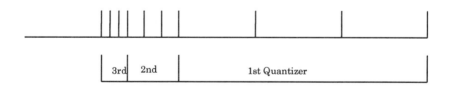

Figure 3-10 *Piecewise linear A/D Converter.*

shape the signal, and then A/D convert it to the digital domain, where the remaining processing is digital. This enables the receiver to be configured for many receive applications and configurations for different waveforms by simply changing the software that drives the digital receiver. The digital signal processing can do everything that the analog processing can do, with the only limitation being the processor throughput required to sample fast enough for the frequency content of the signal.

3.22 Summary

The receiver is an important element in the transceiver design. The receiver accepts the signal in space from the transmitter and amplifies the signal to a level necessary for detection. The superheterodyne is the most used receiver type and provides the most versatility by being able to apply a common IF frequency. Saturation, compression, sensitivity, dynamic range, reduction in unwanted spurious signals and maximizing the S/N are the main concerns in designing the receiver.

3.23 References

[1] James Bao-yen Tsui, *Microwave Receivers with Electronic Warfare Applications*, John Wiley & Sons, New York, 1986.

[2] Curtis M. Abrahamson, *"Phase Noise Seminar"*, Oct. 1984.

PROBLEMS

1. What is placed in the transceiver that prevents the transmitted signal from entering the receiver on transmit?

2. What is the gain required by the receiver for a system that has an 8-bit A/D converter, maximum of 1 volt, NF of 3 dB, a bandwidth of 10 MHz, and source noise of 4 dB?

3. What is the third order, spurious-free dynamic range of the system in problem 2 with a +20 dBm IP3?

4. What is the usable dynamic range of the system in problems 2 and 3 above with a required S/N of 10 dB?

5. Find the NF in 4 above? Find the NF w/o a LNA with an IF NF of 3 dB and the total loss before the IF is 10 dB?

6. Calculate all intermodulation products up to 3rd order for 10 MHz and 12 MHz signals.

7. What is the shape factor for a filter with 0dBm at 90 MHz, −3 dBm at 100 MHz, and −60 dBm at 120 MHz?

8. Find the required sample rate if the highest frequency component is 1 MHz to satisfy the Nyquist criteria?

9. What would be the maximum amplitude and phase error on a linear A/D converter using 2 bits of resolution?

10. What are the advantages and disadvantages for oversampling a received signal?

4

AGC Design And PLL Comparison

Automatic gain control (AGC) is used in a receiver to vary the gain in order to increase the dynamic range of the system. AGC also helps to deliver a constant amplitude signal to the detectors with different amplitude RF signal inputs to the receiver. AGC can be implemented in the RF section, the IF section, or in both the RF and IF portions of the receiver. Most often it is placed in the IF section of the receiver but placement is dependent on the portion of the receiver that limits the dynamic range. The detection of the signal level is usually done in the IF section before the analog/digital (A/D) converter or analog detection circuits. Often the detection is done in the digital signal processing circuitry and is fed back to the analog gain control block. The phase lock loop (PLL) is analyzed and compared to the AGC analysis since both processes incorporate feedback techniques which can be evaluated using control system theory. The similarities and differences are discussed in the analysis. The PLL is characterized only for tracking conditions and not for capturing the frequency or when the PLL is unlocked.

4.1 AGC Design

The RF signal is downconverted, amplified and the output is split and detected for use in the automatic gain control (AGC) to adjust the gain in the IF amplifiers. A voltage controlled attenuator or variable gain amplifier plus a linearizer may be used for the actual control of the amplitude of the signal. The AGC voltage can be used to display the received power within 0.5 dB accuracy by translating AGC volts to a received power equivalent and driving a display or monitoring the level by using a personal computer (PC).

The requirements for AGC in a system are established by using several parameters to determine the amount of expected power fluctuations that might occur during normal operation. Some of the parameters that are generally considered include the variation in distance of operation, propagation loss variations (weather, fading, etc.), multipath fluctuations, variations in the antenna pattern producing a variation in the gain/loss of the antenna in comparison to a true omni antenna, and expected power fluctuations due to hardware variations in both the transmitter and the receiver. The following demonstrates a practical design for an AGC circuit.

A typical automatic gain control used in receivers consists of an AGC amplifier and a feedback voltage to control the gain. A voltage controlled attenuator can be used in place of the AGC amplifier, however, caution must be taken here since the noise figure of the receiver can be changed if the attenuation is large. This feedback system can be designed

AGC Design and PLL Comparison

using basic control system theory. The basic model is shown in Figure 1a.

This model is set up with the maximum gain from the AGC amplifier taken out of the loop as a constant and the feedback loop subtracts gain from this constant depending on the detected signal level.

The feedback system is redrawn to show the point at which the analysis is done for stability as shown in Figure 1b.

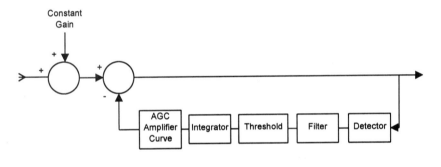

a. Control system for the AGC.

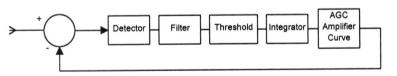

b. Control system analysis for the second order AGC closed loop response.

Figure 4-1 *AGC block diagrams used in the analysis.*

This determines the stability of the AGC loop and the feedback gain is then equal to unity.

4.2 AGC Amplifier Curve

The first step in designing an AGC is to determine the amount of gain control needed. This may be given directly, or calculated given the receiver's range of operation. Sometimes a compression ratio (range of signal in vs. range of signal out) is given. This will determine which AGC amplifier is needed.

Once the AGC amplifier is selected, the gain (dB) vs. control-voltage(V) slope of the AGC amplifier is determined. The best way to do this is actual measurements in the lab. Some companies will furnish this curve. This slope is generally nonlinear. The AGC amplifier should be chosen so that gain vs. control voltage is as linear as possible so that loop gain and noise bandwidth do not change as the control voltage is changed. An estimated linear slope can be applied depending on the design criteria. An example of an amplifier curve and an estimated linear slope is shown in Figure 4-2.

The slope is in dB/V and is used in the loop gain analysis. The estimated linear slope should be chosen where the AGC is operating most of the time. However, the AGC system should be designed so that it does not become unstable across the entire range of operation. The gain constant may be chosen at the steepest portion of slope so the loop gain will always be less than this gain value. This

AGC Design and PLL Comparison

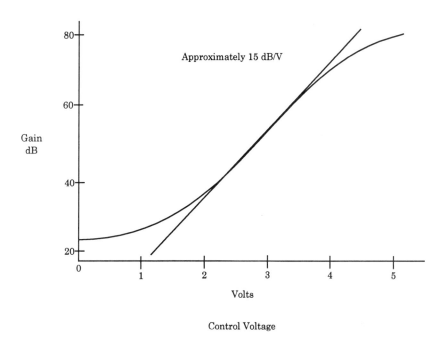

Figure 4-2 *Typical gain vs. control voltage for an AGC amplifier.*

will cause the response time of the AGC to be longer (smaller bandwidth) which means the noise in the loop will always be less. Therefore, since the bandwidth can only get smaller, the reduction of desired modulation will always be smaller. Unless a slower response time presents a problem
(as in the case of fading when a very slow AGC is used), many situations will be able to use this design criteria.

4.3 Linearizer

A linearizer can be used to compensate for the amplifier's non-linearity and create a more constant loop gain response over the range of the amplifier. A linearizer is a circuit that produces a curve response that compensates the nonlinear curve such that the sum of the two curves results in a linear curve (see Figure 4-3). (Note that linear response is on a linear(V) vs. log(P) scale).

A linearizer can be designed by switching in diodes in a piece-wise fashion which alters the gain slope to compensate for the AGC amplifier curve. As the diodes are switched on, they provide a natural smoothing of the piece-wise slopes due to the slope of the diode curve itself. This allows for fairly accurate curve estimations. Note that in Figure 4-3 the sign of the slopes have been neglected for simplicity. Obviously, care needs to be taken to ensure the correct sense of the loop (i.e. negative feedback).

4.4 Detector

The next step is the detector portion of the AGC. The sensitivity (gain) of the detector along with linearity of the curve (approaching a log of the magnitude-squared device) are important parameters in selecting the detector. The linear section of the detector can be chosen on the 'square-law' portion of the curve (low level signal) or the 'linear' portion of the curve (high level signal). The output power or voltage of the AGC is given or chosen for a particular receiver. The operating point can influence the choice of

AGC Design and PLL Comparison 145

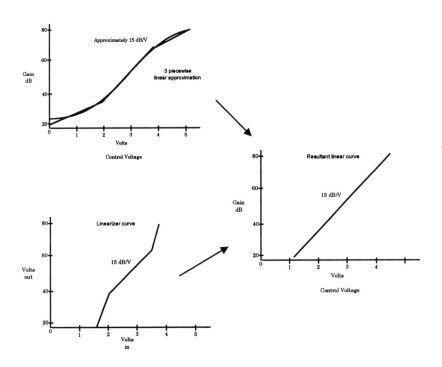

Figure 4-3 *Using a linearizer curve to obtain a linear response for the amplifier control.*

the detector used. The detector provides a voltage-out(V)/power-in(dBm) curve or V/dB slope that can be measured or given. (Note that the power-in is an actual power level and is given in dBm. The change of power-in levels is a gain or loss given in dB. Therefore the slope is a change of power-in levels and is given as V/dB not V/dBm). The ideal detector needs to be linear with respect to:

$$V_o = \log[V_i^2] \text{ (log square law slope).} \quad 4.1$$

The slope can be calculated if the detector is ideal:

$$(V+dV)-(V-dV)/[2\log(V+dV)/V-2\log(V-dV)/V] \quad 4.2$$

where:

V = operating voltage level of the detector.
dV = small variation of the operating voltage level.

Note that if the detector is linear with respect to:

$$V_o = \log[V_i] \text{ (log slope)} \quad 4.3$$

The difference in the slope is a scale factor (2). An amplifier stage with a gain of two following this detector would make it equal to the previous detector. A linearizer could be designed to compensate for either slope.

The slope of a real detector is not linear. However, the slope is chosen so that it is linear in the region where it is

AGC Design and PLL Comparison

operating most of the time (see Figure 4-4). This provides another part of the loop gain given in V/dB.

The non-linearity of the detector will change the instantaneous loop gain of the feedback system. However, unless the system goes unstable, this is generally not a problem since the operating level to the diode is held

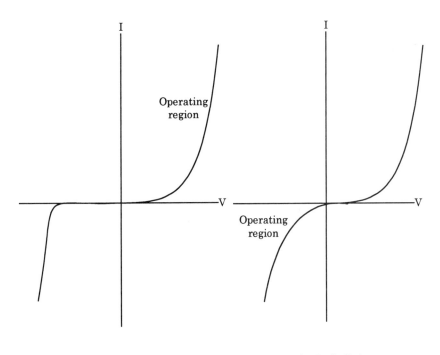

a. Standard diode.   b. Back diode.

Figure 4-4 *Slope of some typical detectors.*

constant by the AGC. Since the diode will always be operating steady state at one point on the curve, the importance of linearization is not significant compared to the AGC amplifier linearization. A linearizer could be used to help create a more linear slope from the detector so that the loop gain is more constant. If the modulating signal is large compared to the non-linearity of the diode curve, there could be some distortion of the modulating signal. Also, there might be some distortion when the signal is on the edge of the AGC range of operation.

The RC time constant following the diode should be much larger than the period of the carrier frequency and much smaller than the period of any desired modulating signal. If 1/RC of the diode detector approaches the loop time response, then it needs to be included in the loop analysis. If 1/RC is greater than an order of magnitude of the loop frequency response, then it can be ignored, which is generally the case.

Since in general, an AGC is not attempting to recover a modulating signal, as in AM detection, failure-to-follow distortion is not considered. However, if the rate of change in amplitude expected is known, then this can be used as the modulating signal in designing for the response time. If a modulating signal (like a conical scan system produces to achieve angle information) is present, the total AGC bandwidth should be at least 10 times smaller to prevent the AGC from substantially reducing the conical scan modulating frequency. Here again, this depends on the design constraints. The modulation is actually reduced by

AGC Design and PLL Comparison 149

1/(1+Loop Gain). This is the sensitivity of the loop to the frequency change of the input.

A back diode provides very good sensitivity as a detector and uses the negative (back biased) portion of the curve for detection (see Figure 4-4). This detector operates well in the presence of low level signals and requires no bias voltage for the diode. The amplifier before the detector needs to be capable of driving the detector. A buffer amplifier could be used to supply the necessary current. Also, the buffer amplifier isolates the received signal from loading effects of the detector. With a power divider and high frequency detectors, this may not be a problem. The detector is matched to the system and the AGC amplifier output can drive the detector directly through the power divider.

4.5 Loop Filter

A loop filter needs to be designed to establish the frequency response of the loop and to stabilize the loop. The loop filter should follow the detector to prevent any high frequency components from affecting the rest of the loop. A passive phase-lag network (lag-lead according to Gardner) is used. The schematic consists of two resistors and a capacitor and is shown in Figure 4-5 along with the transfer function. This filter provides a pole and a zero to improve stability of the loop. This filter is used to affect the attenuation and the roll-off point since the phase shift for a phase lag filter generally has a de-stabilizing effect.

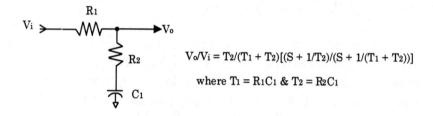

$V_o/V_i = T_2/(T_1 + T_2)[(S + 1/T_2)/(S + 1/(T_1 + T_2))]$

where $T_1 = R_1C_1$ & $T_2 = R_2C_1$

a. This is a lag filter providing a zero and a pole.

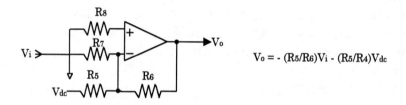

$V_o = - (R_5/R_6)V_i - (R_5/R_4)V_{dc}$

b. This is a threshold amplifier summing in the offset voltage which determines the AGC output level.

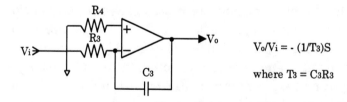

$V_o/V_i = - (1/T_3)S$

where $T_3 = C_3R_3$

c. This is an integrator to make the system a type 1 feedback system.

Figure 4-5 *Loop filter components.*

AGC Design and PLL Comparison

4.6 Threshold Level

A threshold needs to be set to determine the power level output of the AGC amplifier. A voltage level threshold in the feedback loop is set by comparing a stable voltage level with the output of the loop filter. This sets the power level out of the AGC amplifier. A threshold level circuit is shown in Figure 4-5 and consists of an operational amplifier with a DC offset summed in. For this example, if the output of the loop filter is less than the offset, then the output of the threshold device will be negative. If the loop filter output is greater than the offset, the output will be positive. This determines which way the gain is adjusted. The offset is equal to $-(R5/R4)V_{dc}$ as shown in Figure 4-5.

An additional gain is added to the loop and is selected for the particular AGC signal level and diode gain. This gain should be large enough to increase threshold sensitivity but small enough to prevent saturation of the loop amplifiers. The gain constant is defined as K_c and is equal to $-(R5/R6)V_1$, see Figure 4-5.

4.7 Integrator

With just the phase-lag filter in the loop, the system is classified as a type 0 system and there will be a steady-state error for a step response. If an integrator is included in the feedback loop, then the system is a type 1 system and the steady state error for a step response is zero. Therefore an integrator is included in the loop. An integrator, using an operational amplifier, is shown in Figure 4-5. The output of the threshold device is integrated before controlling the gain of the AGC amplifier. With no signal present, the integrator will integrate to the rail of the operational amplifier (op amp) and stay there until a signal greater than the threshold is present. The AGC amplifier is held at maximum gain during this period of time which is the desired setting for no signal or very small signals. The response time for the op amps to come out of saturation is negligible compared to the response time of the loop. If the voltage level out of the loop amplifiers is too large for the AGC amplifier, then a voltage divider should be used or a voltage clamp to protect the AGC amplifier. Do not limit the DC gain of the integrator or there will be a steady-state error. This makes the loop a second order system and is easily characterized by using control system theory.

4.8 Control Theory Analysis

The open loop transfer function including the integrator is:

$$Tsol = \frac{K_a K_d K_c T_2 (S + \frac{1}{T_2})}{T_3 (T_1 + T_2) S (S + \frac{1}{T_1 + T_2})} \quad 4.4$$

where:

AGC Design and PLL Comparison

K_a = AGC amplifier loop gain.
K_d = detector loop gain.
K_c = gain constant
T_1 = R1×C1
T_2 = R2×C1
T_3 = R3×C3
T_{sol} = open loop transfer function.
$S = j\omega$ (assumes no loss)

The closed loop response is:

$$Tscl = \frac{K(S + \frac{1}{T_2})}{S_2 + (\frac{1}{T_1 + T_2} + K)S + \frac{K}{T_2}}$$

4.5

where:

$K = K_a K_d K_c (1/T_3)(T_2/[T_1+T_2])$
T_{scl} = closed loop transfer function.

From control system theory for the second order system above:

$$2z\omega_n = K + 1/(T_1+T_2)$$

4.6

and:

$$\omega_{n2} = K/T_2$$ 4.7

where:

ω_n = natural frequency of the system.

z = damping ratio.

The natural frequency ω_n should be set at least 10 times smaller than any desired modulating frequency. The damping ratio is chosen so the system responds to a step function as fast as possible within a given percent of overshoot. With a 5 percent overshoot, the minimum damping ratio is .707. This means the system is slightly underdamped. The damping ratio is chosen depending on design criteria. If the overshoot is too high, the damping ratio should be changed. Once the damping ratio and the natural frequency are chosen for a system, the time constants can be mathematically solved.

The poles and zeros for the root locus criteria are found by using G(s)H(s). Note that $1/T_2$ determines the zero location for both the open loop and closed loop cases since H(s) = 1. The zero is placed at $-2z\omega_n$ and the root locus plot is shown in Figure 4-6.

This moves the root locus, the migration of the poles, away from the jω axis to prevent oscillatory responses with change in gain. This also places the poles on the tangent of the root locus circle at the given damping ratio so that as the gain varies the damping ratio is always greater. One pole, due to the integrator, migrates from the jω axis along the sigma axis as the loop gain constant, K increases. As K approaches zero, the pole due to the integrator lies on

AGC Design and PLL Comparison

the jω axis. This is something to keep in mind when designing with nonlinear devices. As the loop gain decreases, the integrator pole migrates closer to the jω axis and becomes more oscillatory. However, a pole on the axis

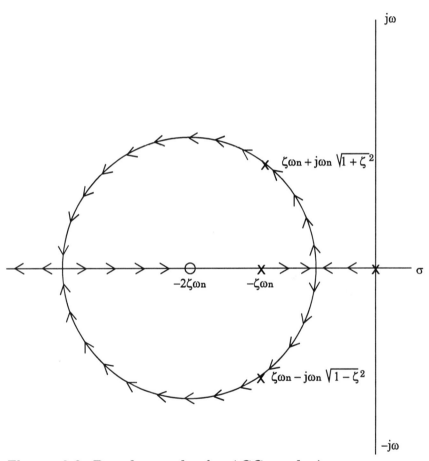

Figure 4-6 *Root locus plot for AGC analysis.*

means that the system is over-damped [z>1] and will not oscillate. The only way this system can become unstable is with stray poles close to the origin which is unlikely. The loop gain K is chosen in this example so the operating point lies on the root locus circle with the desired damping factor (see Figure 4-6).

$$T2 = 1/2z\omega_n$$
$$T1 = 1/(2z\omega_n - K) - T2$$
$$K = K_a K_d K_c (1/T3)(T2/[T1+T2])$$
$$T3 = K_a K_d K_c (T2/[T1+T2])/K \qquad 4.8$$

Once the time constants are solved, the selection of the resistors and capacitors is the next step. Low leakage capacitors are preferred and the resistors should be kept small. This is to prevent current limiting of the source by keeping the current large compared to the current leakage of the capacitor. There are tradeoffs here depending on parts available and values needed.

$$T1 = R1C1 \quad R1 = T1/C1 = .0707/1\mu F = 70.7 k\Omega$$

$$T2 = R2C1 \quad R2 = T2/C1 = .0707/1\mu F = 70.7 k\Omega$$

$$T3 = R3C3 \quad R3 = T3/C3 = 1.0/1\mu F = 1 M\Omega$$

4.9 Modulation Frequency Distortion

The reduction of the modulating frequency (con-scan frequency information) is calculated by solving the closed-loop transfer function for the modulating frequency ω_m.

AGC Design and PLL Comparison

The term magnitude means that for an input gain change of a signal level at the modulating frequency ω_m, the AGC amplifier will change its gain by the magnitude of the input gain change. The modulating signal is reduced by the magnitude times the input gain change.

For example, if the input signal level changes from 0 dBm to 10 dBm at the modulating frequency, the input gain change is 10 dB. The feedback loop response reduces the 10 dB change by the magnitude of say 0.04 or 0.4 dB. This means the AGC amplifier changes gain by 0.4 dB for an input change of 10 dB. However, since there is a phase shift of the feedback signal, the amplitude reduction will be less since it is associated with the cosine of the angle.

To put it in equation form:

$$\text{Signal input} = P_{in}[dBm] + A\cos(\omega_m t)[dB] \quad 4.9$$

$$\text{Signal output} = P_{in}[dBm] + A\cos(\omega_m t)[dB] - HA\cos(\omega_m t + C)[dB] \quad 4.10$$

where:
  A = amplitude of the modulating signal (con-scan).
  H = amplitude through the feedback loop.

  C = phase shift through the loop.
  t = time

The superposition of the two cosine waves in the output equation result in the final gain in dB that is added to the power. Another way to look at this is to consider taking the change in dB caused by the input modulation amplitude and phase, subtracting the loop input and phase by converting to rectangular coordinates, and then converting back to polar coordinates. For example:

Input = 10 dB   angle = 0°
Loop response = mag (.04×10) = .4 dB   angle (–90°+0°) = –90
Total response = (10+j0)–(0–j.4) = 10+j.4 = 10.008 at an angle of 2.3°.

The magnitude changed by 0.008 dB, and the phase of the modulating signal shifted by 2.3 degrees.

Note that this assumes that the change in gain is within the dynamic range of the AGC. If the gain change is outside of the AGC dynamic range, then the magnitude of the loop response is 0.04 times only the portion of AGC range that is being used.

To determine the resultant amplitude and phase for the general case:

$$H(\omega) = [H]e^{jC} = [H]\cos C + j[H]\sin C \qquad 4.11$$

so:

$$\{1-H(\omega)\} = 1-\{[H]\cos C - j[H]\sin C\} \qquad 4.12$$

AGC Design and PLL Comparison

The magnitude equals:

$$[1-H(\omega)] = \sqrt{\{1+[H]^2-2[H]\cos C\}} \qquad 4.13$$

and the angle equals:

$$\tan^{-1}\{-[H]\sin C/(1-[H]\cos C)\} \qquad 4.14$$

4.10 Comparison of the PLL and AGC Using Feedback Analysis Techniques

Phase lock loops (PLLs) are important elements in the design of synthesizers and carrier recovery circuits. They are used in almost all receiver designs and have become a basic building block design tool for the engineer. This article simplifies the analysis of the stability of the PLL and makes a comparison between automatic gain control (AGC) and the PLL.

4.11 Basic PLL

The basic PLL standard block diagram is shown in Figure 4-7 which should be familiar to most engineers. The operation of the PLL takes the input frequency and multiplies this frequency by the voltage controlled oscillator (VCO) output frequency and if the frequencies are the same, then a phase error is produced. The nature of the PLL forces the phase of the VCO to be in quadrature with the phase of the incoming signal. This produces the sine of the phase error:

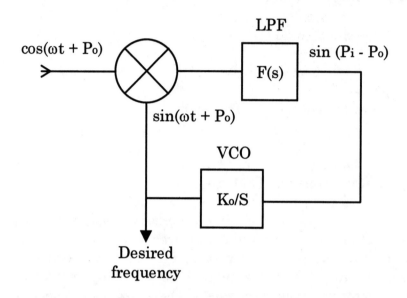

Figure 4-7 *Basic diagram of a phase lock loop.*

$$\cos(\omega t+a)\sin(\omega t+b) = 1/2[\sin(a-b) + \sin(2\omega t+a+b)]$$

Using a low pass filter to eliminate the sum term produces:

$$= 1/2[\sin(a-b)] \qquad 4.15$$

if a = b then:

$$= 1/2[\sin(0)] = 0.$$

AGC Design and PLL Comparison

The zero voltage output sets the VCO to the correct steady-state value. If the voltage changes due to a change in input phase, the VCO changes until the phase error is approximately equal to zero. The low pass filter removes higher frequency components and also helps to establish the loop gain. The PLL tracks a phase change over time and since the change of phase per unit time is frequency, the PLL tracks frequency. Since it performs this task by converting the changing phase into a changing voltage that controls the VCO, the analysis is performed using phase as the parameter.

4.12 Comparisons of the PLL and AGC

Phase lock loops can be designed using basic control system theory since it is a feedback system. The analysis is very closely related to the automatic gain control study which was performed previously [1]. In order to analyze the stability of the PLL, the basic diagram is redrawn (see Figure 4-8).

The block diagram for the PLL is almost identical to the AGC block diagram provided in Figure 4-1. The VCO contains an integrator and a gain constant, whereas the AGC amplifier only has a gain constant and the integrator is added as another block. The integrator included in the VCO produces a type 1 system, so for a step change in phase the PLL will have a steady-state error of zero. For the AGC, the added integrator provided a steady state error of zero for a step change (in dB) of input power. If a zero steady state error for a step in frequency is desired,

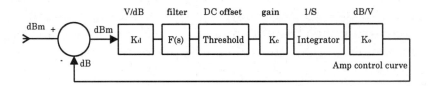

a. AGC control system analysis block diagram.

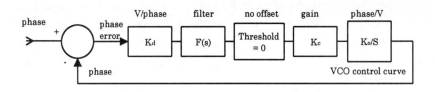

b. PLL control system analysis block diagram.

Figure 4-8 *Comparison between AGC and PLL feedback systems.*

AGC Design and PLL Comparison

another integrator needs to be included in the PLL. However, this chapter will limit the analysis to a second order, type 1 system.

4.13 Detector

The phase detector in the PLL performs two operations, see Figure 4-8. First, the detector takes the input phase and subtracts the feedback phase from the VCO to produce a phase error. Second, the detector converts the phase into voltage with a particular slope depending on the phase difference and the frequency of operation. This process for the AGC used power in dBm for the input and subtracted the gain in dB to produce a power level error which was converted to voltage. The AGC detector's operation point was set at the desired dBm output so the detector gain was the same around the operation point regardless of the input operating power. If the desired frequency is set for the PLL, then the same criteria applies. However, in many cases when a PLL is used, the desired operating frequency changes (such as the case for synthesizer design). Therefore, the detector gain becomes a function of the input frequency or the input operating point. Care must be taken when selecting the gain constant for the VCO detector because it depends on the input frequency. There are approaches to linearizing the detector slope [2], but for this analysis the slope is chosen as a constant. If the frequency is constant, then the slope is easily chosen and the linearity is generally not a problem.

4.14 Loop Filter

The passive phase-lag network used for the loop stabilizing filter is identical for both the AGC and the PLL. The actual numerical values will be different depending on the loop requirements. The loop filter should follow the detector to prevent any high frequency components from affecting the rest of the loop. The schematic consists of two resistors and a capacitor and is shown in Figure 4-5 along with the transfer function. This circuit provides a pole and a zero to improve stability of the loop. The filter is used to affect the attenuation and the roll-off point since the phase shift for a phase-lag network generally has a destabilizing effect.

4.15 Loop Gain Constant

An additional gain is added to the loop and is selected for the particular operating levels of the PLL. The gain constant is defined as K_c and is equal to $-R5/R6$, see Figure 4-5. The gain constant is used in both the PLL and the AGC cases and are different depending on each of their proper operating points. An added threshold is not required in the PLL as was in the AGC since the desired phase difference is always zero regardless of the desired frequency of operation.

AGC Design and PLL Comparison   165

4.16 Integrator

As mentioned previously, the integrator for the PLL is inherent in the loop whereas the AGC integrator is added to the loop. The reason for this is because the PLL operates using phase as the parameter and the VCO actually delivers frequency. Therefore, the PLL forces the VCO to deliver phase which inherently adds an integrator since the integral of frequency is phase.

4.17 Conversion Gain Constant

The conversion in the AGC circuitry from volts to dBm produces a slope gain constant (dB/volts) and is chosen considering the tradeoff between stability and response time. The conversion in the PLL is actually in two steps. First, the slope conversion constant converts from voltage to frequency which is the constant for the VCO labeled K_o. Second, the inherent integration to convert frequency into phase. The constant, K_o, for the PLL is not linear. This was also true for K_a in the AGC analysis, so a linear approximation is used. A linearizer could be implemented for either the PLL or the AGC circuits.

4.18 Control Theory Analysis

The open loop transfer function for the PLL is identical to the open loop transfer function for the AGC except that the integrator is considered as $1/S$ instead of $1/(T_3 S)$ respectively as shown below:

$$Tsol = \frac{K_a K_d K_c T_2 (S + \frac{1}{T_2})}{T_3(T_1 + T_2) S (S + \frac{1}{T_1 + T_2})} \qquad 4.16$$

$$Tsol = \frac{K_o K_d K_c T_2 (S + \frac{1}{T_2})}{(T_1 + T_2) S (S + \frac{1}{T_1 + T_2})} \qquad 4.17$$

where:

K_a = AGC amplifier loop gain.
K_o = VCO gain.
K_d = Detector loop gain.
K_c = gain constant
T_1 = R1×C1
T_2 = R2×C1
T_3 = R3×C3
T_{sol} = open loop transfer function.
S = jω (for a lossless system)

The closed loop response is:

AGC Design and PLL Comparison

$$Tscl = \frac{K(S+\frac{1}{T_2})}{S^2 + (\frac{1}{T_1+T_2}+K)S + \frac{K}{T_2}}$$

4.18

where:

$K = K_a K_d K_c (1/T_3)(T_2/[T_1+T_2])$ for the AGC.
$K = K_o K_d K_c (T_2/[T_1+T_2])$ for the PLL.
T_{scl} = closed loop transfer function for both the PLL and AGC.

From control system theory for the second order system above:

$$2z\omega_n = K + 1/(T_1+T_2) \quad 4.19$$

and:

$$\omega_{n2} = K/T_2$$

where:

ω_n = natural frequency of the system.

z = damping ratio.

This analysis applies to the condition of the loop being already phased locked and not on capturing the signal.

The loop gain for the PLL is selected considering several factors:

(1) Stability of the loop.
(2) Lock range and capture range.
(3) Noise in the loop.
(4) Tracking error.

The wider the bandwidth, the larger the lock and capture ranges and also the more noise in the loop. The lock range is equal to the DC loop gain. The capture range increases with loop gain and is always less than the lock range. If the maximum frequency deviation or the maximum phase error desired is given, then this establishes the open loop gain. When designing a synthesizer, the crossover between the crystal reference and the VCO with respect to phase noise is chosen for best noise performance for the loop bandwidth. Further study is suggested since this chapter only deals with the stability of the PLL in the locked condition. Once the frequency is chosen, the same stability analysis for the PLL can be performed as was done for the AGC circuit analysis.

The damping ratio is chosen so the system responds to a step function as fast as possible within a given percent of overshoot. With a 5 percent overshoot, the minimum damping ratio is .707. This means the system is slightly underdamped. The damping ratio is chosen depending on design criteria. If the overshoot is too high, the damping

AGC Design and PLL Comparison

ratio should be changed. Once the damping ratio and the natural frequency are chosen for a system, the time constants can be mathematically solved. An example is shown below:

Determine loop bandwidth = 160 Hz

Choose ω_n = 1.0krad/sec. f_n = 160 Hz
Choose z = .707   P.O. = 5%

The poles and zeros for the root locus criteria are found by using G(s)H(s). Note that $1/T_2$ determines the zero location for both the open loop and closed loop cases since H(s) = 1. The zero is placed at $-2z\omega_n$ and the root locus plot is shown in Figure 4-6. This moves the root locus, the migration of the poles, away from the jω axis to prevent oscillatory responses with change in gain. This also places the poles on the tangent of the root locus circle at the given damping ratio so that as the gain varies the damping ratio is always greater. One pole, due to the integrator, migrates from the jω axis along the sigma axis as the loop gain constant, K increases. As K approaches zero, the pole due to the integrator lies on the jω axis. This is something to keep in mind when designing with nonlinear devices. As the loop gain decreases, the integrator pole migrates closer to the jω axis and becomes more oscillatory. However, a pole on the axis means that the system is over-damped [z>1] and will not oscillate. The only way this system can become unstable is with stray poles close to the origin. The loop gain K is chosen in this example so the operating point lies on the root locus circle with the desired damping factor see Figure 4-6.

For the example:

Choose T_2:

$$T_2 = 1/2z\omega_n = 1/1.414k = 707.2 \times 10^{-6}. \quad 4.20$$

Using the root locus equations above:

$$K = T_2\omega_{n2} \quad 4.21$$

$$2z\omega_n = T_2\omega_{n2} + 1/(T_1+T_2) \quad 4.22$$

And solving for T_1:

$$T_1 = [1/(2z\omega_n - T_2\omega_{n2})] - T_2 \quad 4.23$$

$$T_1 = [1/\{1.414k - (707.2\times 10^{-6})1000k\}] - 707.2\times 10^{-6} = 707.6\times 10^{-6}$$

For the AGC case the constants are specified and T_3 is solved as shown:

$$K = K_a K_d K_c (1/T_3)(T_2/[T_1+T_2]) \quad 4.24$$

$$T_3 = K_a K_d K_c (T_2/[T_1+T_2])/K \quad 4.25$$

$$K = T_2\omega_{n2} \quad 4.26$$

$$T_3 = K_a K_d K_c (T_2/[T_1+T_2])/(T_2\omega_{n2}) \quad 4.27$$

For the PLL case, K_o and K_d are specified and K_c is solved for as shown:

AGC Design and PLL Comparison

$$K = K_o K_d K_c (T_2/[T_1+T_2]) \qquad 4.28$$

$$K_o = 20 \quad K_d = .90$$

$$K_c = K/[K_o K_d (T_2/[T_1+T_2])] \qquad 4.29$$

$$K = T_2 \omega_{n2} \qquad 4.30$$

$$K_c = T_2 w_{n2}/[K_o K_d (T_2/[T_1+T_2])] \qquad 4.31$$

$=707.2\times10^{-6}(1.0\times10^6)/[20(.9)(707.2\times10^{-6})/(707.6\times10^{-6}+707.2\times10^{-6})] = 78.6$.

To complete the design:

Choose C1 = .01 µF

T_1 = R1C1 R1 = T_1/C1 = 707.6×10^{-6}/.01µF = 70.76 kΩ

T_2 = R2C1 R2 = T_2/C1 = 707.2×10^{-6}/.01 µF = 70.72 kΩ

The resultant transfer functions are:

*T(open loop) = [707/S][S+1.414k]/[S+1.403k] 4.32

*T(closedloop)=707[S+1.414k]/[S^2+2.11kS+1000k] 4.33

4.19 Summary

AGC is an important element in the design of all types of transceivers. AGC provides the necessary dynamic range required for operation over varying distances and conditions. AGC can be analyzed using control system theory since it is a feedback system. The similarities between AGC and the PLL are remarkable. They both can be analyzed using modern control system techniques. The steady state error for both systems is zero for a step response. The step response for AGC is related to a power level and the step response for the PLL is for a phase change. If a zero steady state error is required for a step in frequency, then another integrator needs to be added. Further study of the PLL is required to answer various questions such as capture and lock range, operation of the loop in a no-lock situation, and considerations concerning stray poles which may change the simple analysis. However, the idea of using these control system techniques for analyzing both AGC and the PLL can be useful.

4.20 References

[1] R. C. Dorf, *Modern Control Systems*, Addison-Wesley, 1974.

[2] F. M. Gardner, *Phaselock Techniques*, John Wiley & Sons, New York, 1979.

[3] Alan B. Grebene, "The monolithic phase-locked loop--a versatile building block", *IEEE Spectrum* March 1971.

[4] S.R. Bullock, D. Ovard, "Simple Technique Yields Errorless AGC Systems", *Microwaves and RF*, Aug. 1989.

[5] Scott R. Bullock, "Control Theory Analyzes Phase-Locked Loops", *Microwaves and RF*, May 1992.

Problems

1. Name at least 2 RF devices that can be used for gain control.

2. What would be a reasonable value of the RC time constant, following the diode detector for the AGC, be for a carrier frequency of 10 MHz and a desired modulating frequency of 1 MHz?

3. What does an integrator in the feedback loop do for the steady state error?

4. Explain how the integrator achieves the steady state error in problem 3 above.

5. What is the effect if a resistor is placed in (a) series with the capacitor in an op amp integrator, (b) parallel with the capacitor in an op amp integrator?

6. How does a non-linear function affect the response of the AGC loop?

7. Why is a diode curve approximation much better that a piecewise linear approximation?

8. What potential problem does a PLL have in the analysis with infinite gain applied as in the case of an integrator and possible DC offsets with no signal applied that the AGC does not?

AGC Design and PLL Comparison 175

9. What is the general reason that the AGC analysis is similar to the PLL analysis?

10. What state is the PLL assumed to be in the AGC/PLL analysis in this chapter?

5

Demodulation

The demodulation process takes the received signal and recovers the information that was sent. This is the key element in the transceiver process. Demodulation provides the tracking loops and detection processes, and requires careful design to ensure maximum likelihood of retrieving the data with the minimum bit error rate (BER). There are basically two ways to analyze the demodulation process of digital modulated signals for phase shift keying (PSK) signals depending on the type of modulation waveform used and the corresponding demodulation process used. The first way is using an asynchronous process that combines the phase shift times (bits or chips) using a matched filter. This matched filter generally consists of a tapped delay line with the tap spacing equal to the bit or chip duration and provides a means to sum all of the phase shifts. This can be done either using analog tapped delay lines or digital filters. Most systems use digital means for implementing this type of demodulation scheme. The second way to analyze demodulation of a PSK signal is to coherently demodulate the incoming signal using several methods of implementation of various tracking loops. Since PSK systems suppress the carrier, a carrier recovery loop is used to re-establish the carrier to down convert the signal to baseband. Also, if spread spectrum is used, there needs to be a way of stripping off the spread spectrum code to obtain the data stream. This is usually done with a sliding correlator. Once the data stream is produced, there needs to be a way to sample the bits to recover the digital data that was sent. This is usually accomplished by using a bit synchronizer or just bit synch to ensure that the system samples the data in the correct place, which is the center of the bit that was sent. This is the point with the best S/N in what is known as the eye pattern. These methods of demodulation are discussed in detail in this chapter.

5.1 Pulsed Matched Filter

The matched filter process is used for asynchronous detection of a spread spectrum signal. It can be used for pulsed systems where the arrival of the pulse provides the information for the system. The time of arrival (TOA) is used in conjunction with a pulse position modulation (PPM) scheme to decode the data that was sent. This is a very useful for systems that do not transmit continuous data and that do not want to use extra bits (overhead bits) for synchronizing the tracking loops in a coherent system. This type of matched filter is very simple but requires a considerable amount of hardware. This process sums the pseudo-noise (PN) signals in such a way that the signal is enhanced, however, the bandwidth is not changed. An example of this type of matched filter correlator is an acoustic charge transport (ACT) device. This device separates packets of sampled analog data with respect to delay and weights each packet, and then sums them all

Demodulation

together to receive a desired pulse. This can be done digitally using finite impulse response (FIR) filters with the tap delay equal to the chip width and weights that are 1 or −1 depending on the code. This matched filter correlator provides a means of producing a TOA with a high S/N due to combining all of the pulses that were sent into one time frame.

5.2 Matched Filter Correlator

The matched filter correlator consists of a tapped delay line, with the delay for each of the taps equal to 1/chip rate. Each of the delay outputs are multiplied by a coefficient, usually ± 1, depending on the code that was sent. The values will be the reverse of the PN-code values. For example:

    Code values: 1, −1, 1, 1

    Coefficient values: 1, 1, −1, 1

Therefore, the coefficient values are time reversed from the code.

As the signal processes through the matched filter, the weights will correspond at one code delay in time and the output of the resultants are all summed together to create a large signal (see Figure 5-1).

180   Transceiver System Design

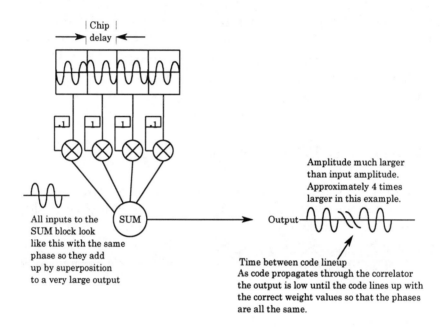

Figure 5-1 *Matched filter analysis.*

If the code is longer and random, then passing a sine wave through it would result in superposition of many segments of the sine wave, non-inverted and inverted so that the net result would be close to 0, or at least reduced. The same thing applies with different codes that are passed through the matched filter since they switch the phase of the carrier at the wrong places and time. The closer the code

Demodulation

is to the pseudo-random code, the higher the output correlation peak will be. Note, however, regardless of the code used, the bandwidth has not changed. The bandwidth is dependent on the chip rate or pulse width which is still 2 times the chip rate or 2/PW for double-sided bandwidth null-to-null see Figure 5-2.

There is no carrier recovery since the process is asynchronous and the signal is therefore demodulated. This method of detection is useful in pulsed systems or time division multiple access (TDMA) systems where the overhead required to synchronize the system would overwhelm the amount of data to be sent. In other words, the overhead bits and time to synchronize the tracking loops reduces the amount of information that can be sent for a given transmission pulse width.

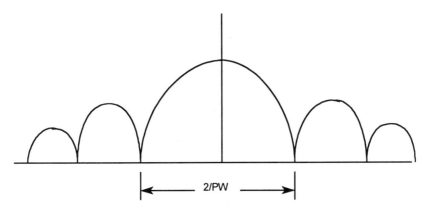

Figure 5-2 Sinc function squared for pulse modulation.

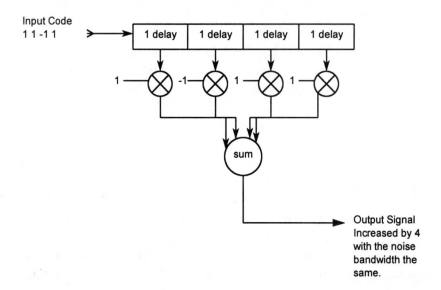

Figure 5-3 *Correlator showing time-reversed algorithm.*

This process can also be done digitally. The signal is quadrature downconverted, which strips the carrier and provides both I and Q digital data streams. The digital quadrature data streams are fed into tapped delay lines where the tap delays are equal to 1 chip period. The outputs of the FIR filter are multiplied by the weights which are the time reversed pattern of the code that was sent. An example is shown in Figure 5-3.

This example shows a code that is 4 bits long. The output of the digital summation is a pulse amplitude of 4 times the input pulse amplitude. This provides the S/N necessary to detect the signal with the noise power being unchanged.

Demodulation 183

Once the pulse is recovered, this produces a pulse in time which establishes the TOA for that pulse. This TOA can be applied to a PPM technique to demodulate the data that was sent.

5.3 Pulse Position Modulation

Pulse position modulation (PPM) measures the TOA of the pulse received to determine where in time the pulse occurred. If absolute time is precisely known, then an absolute PPM demodulation process can be used. Therefore, the TOA reference to absolute time provides the information. The amount of information this TOA pulse provides depends on the PPM grid or the number of possible time divisions that TOA could occur. For example, if there are eight possibilities for the TOA to occur, then there are eight bits of information produced on the TOA of one pulse (see Figure 5-4).

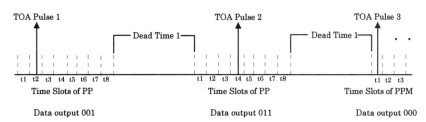

Figure 5-4 *Absolute pulse position modulation for data encoding.*

The PPM grid is referenced to an absolute time mark and the entire PPM grid changes only with the absolute time variations. Therefore, the PPM grid is not dependent on the TOA pulses received. Once the PPM structure is set up, it does not change. The dead time is allocated to ensure that the chips that make up the pulsed output are through the matched filter and do not interfere with other pulses. The dead times are the same lengths and do not vary in length between pulses.

Another scheme that is used when absolute time is not known is differential PPM. This method relies on the difference in time between one received TOA pulse and the next received TOA pulse. Therefore, the first TOA pulse received contains no data but provides the reference for the following TOA pulse received. The time after a received TOA pulse is divided up into a PPM grid and when the following TOA pulse is received in the grid determines the data that was sent. For example, if the time on the grid is divided up into eight time slots after receipt of a TOA pulse, the next TOA pulse falls in one of the eight time slots and produces eight bits of information (see Figure 5-5).

Then the third TOA pulse is mapped into the PPM grid with the second TOA pulse taken as the reference and so on as shown in Figure 5-5. The dead time is also required for this PPM scheme, however, the dead time spacing between the PPM grids is dependent on the location of the received reference pulse. For example, if a TOA pulse is received in time slot 1 of the PPM grid used for the reference of the next TOA pulse, then the dead time is

Demodulation

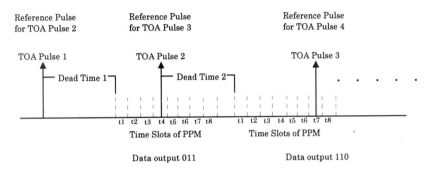

Figure 5-5 *Differential pulse position modulation for data encoding.*

from the time of reception of the reference TOA pulse. If the same TOA pulse is received in time slot 8 of the PPM grid and is used as the reference for the next pulse, the absolute time of the PPM grid for the next pulse will be delayed more than if time slot 1 was used as the reference. However, for the absolute time PPM mentioned previously where the time is known accurately, the PPM grid for all pulses remains the same and is not dependent on the previous pulses received.

5.4 Code Division Encoding

Another type of demodulation scheme using the matched filter concept is to use a code division encoding. This contains a matched filter for every code sequence used. Each of the TOA pulses have different codes dependent on the data sent. For example, if there are 8 different code sets and 8 different matched filters, an output on code 1 decodes to data of 000 indicating that code 1 was sent. If there is an output of the 4th matched filter, then the data decodes to 011 which means that code 4 was sent. The major drawback to this type of demodulation is the amount of hardware required to build multiple matched filters.

5.5 Coherent Demodulation

Coherent demodulation is generally used for continuous systems or long pulse systems that can afford the overhead required to enable the tracking loops in the system. This overhead includes the bits that need to be sent so that the tracking loops can start track the incoming signal. These bits are wasted because the coherent data has not been demodulated since the tracking loops are not tracking the data yet. Therefore, if there is an output, it will contain a tremendous amount of errors. For short pulse systems, this overhead can take up most of the signal. Every time a pulse is sent out, the tracking loops have to re-acquire the signal. There are many ways that have been devised to demodulate the incoming signal and many ways to implement the tracking loops for each part of the system.

The demodulation process requires three basic functions in order to retrieve the data that was sent. They are:

    a.   Remove spreading spectrum coding, if using spread spectrum techniques. This is generally done using a despreading correlator or equivalent.

Demodulation

    b.   Recover the carrier, since the digital modulation results in a suppressed carrier and in order to strip off the carrier one needs to be generated.

    c.   Align and synchronize the sample point for sampling the data stream at the optimal S/N point, which requires a bit-synch.

5.6 Despreading Correlator

The method for removing the spread spectrum code on a spread spectrum system is called a despreading correlator. Despreading the signal allows for a smaller noise bandwidth to be used which decreases the noise power, thus increasing the S/N of the received signal. A sliding correlator incorporates a method of stripping off the PN code to regain the data information. This is generally the first stage in the detection process since the bandwidth is large due to spreading. The carrier frequency needs to be at least two times the bandwidth. This prevents the main lobe and one other lobe from folding over into the signal resulting in aliasing. Since the code can be long before it repeats (some codes do not repeat for 2 years), an initial alignment needs to be done. This requires additional bits to be sent that are not related to the data being sent. These bits are called overhead bits, and will change the data rate by a certain percentage of the data messages sent. For continuous systems, these overhead bits are a very small percentage of the message. For pulsed type systems or TDMA systems, these bits can be a large percentage depending on the length of the pulse. The code alignment process can be accomplished by using the following techniques:

    a.   Generating a short code for acquisition only.

    b.   Using highly reliable clocks available to both the transmitter and receiver.

    c.   Using an auxiliary subsystem that distributes system time to both parties.

In many systems architectures, all of these techniques are used to simplify and accelerate the code alignment process. Once the code time is known fairly accurately, two steps are used to demodulate the PN code. The first step is to determine if the code is lined up and the second step is to lock the code into position. The first step is achieved by switching a non-correlated code into the loop and then into a sample and hold function to measure the correlation level (noise) output. Then the desired code is switched into the loop and into the sample and hold function to measure the correlation level (signal) output. The difference in these levels is determined and the results are then applied to a threshold detector to determine if the search should continue. If the threshold level is achieved, the second step is activated. This prevents false locking of the sidelobes. Note that false lock occurs at one-half of the bit rate. The second step is to maintain lock (also fine tuning) by using a code lock tracking loop. This assumes that the codes are within 1/2 bit time. One type of code tracking loop is

Demodulation

called an early-late gate or Tau-dither loop. An early-late gate switches a VCO between a early code (1/2 bit time early) and a late code (1/2 bit time later). The correlation or multiplication of the early code and the input code, and the late code and the input code produces points on the autocorrelation peak as shown in Figure 5-6.

This shows the code loop locked when the values of the autocorrelation function are equal. Therefore, the non-delayed code is used for demodulation of the spread spectrum waveform. The peak of the autocorrelation function occurs when the non-delayed, aligned code is

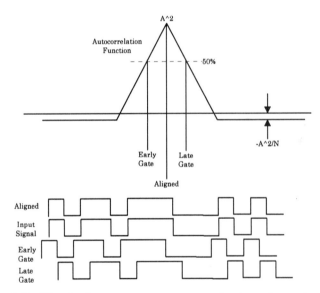

Figure 5-6 *Early-late gate autocorrelation function.*

multiplied with the incoming signal and summed (see Figure 5-6). These codes are mixed with the incoming signal, filtered and detected (square-law envelope detected), and the output sign is changed at the dither rate. For example, the early code detector output is multiplied by a negative one and the late code detector output is multiplied by a positive one and the final output is filtered by the loop filter which takes the average of the two levels. This provides the error and controls the VCO to line up the codes so that the error is zero. This means the VCO is switching symmetrically around the correlation peak.

Another form of a code lock loop is called a delay-lock loop (DLL). The delay-lock loops accomplish the same thing as the early-late gate, that is, aligning up the code in order to strip off the PN code using autocorrelation techniques. The main difference in the DLL compared to the early-late gate is that the DLL uses a non-delayed code and a code that is delayed by one chip. The DLL splits the incoming signal and mixes with VCO output that is either shifted a bit or not, respectively. The main idea is to obtain the maximum correlated signal on the non-delayed code and the minimum correlated signal on the 1 chip delayed code. The DLL generally tracks more accurately (approximately 1 dB) but is more complex to implement.

5.7 Carrier Recovery

Another required function in the coherent demodulation process is to recover the carrier of the incoming signal.

Demodulation

The carrier recovery loop is needed since the carrier in a digitally modulated spread spectrum signal is suppressed. In order to remove the carrier coherently, a carrier recovery tracking loop is used to recover the carrier, and then this recovered carrier is used to strip off or remove the signal's carrier. Generally a VCO is used in the carrier recovery loop for this demodulation process. Once the code and the carrier are removed, only the data remains. There are different methods of performing this task. The most common ways of carrier recovery are:

   a.   Squaring loop
   b.   Costas loop

These loops are discussed in further detail in the paragraphs ahead. There are pros and cons concerning which type of carrier recovery loop should be used. A careful look at these parameters will provide the best type of carrier recovery loop used in any given system design.

5.7.1 Squaring Loop

The squaring loop is used for demodulation of a BPSK direct sequence waveform. The advantage of the squaring loop is mainly the simplicity of implementation. The squaring loop is straight forward and easy to understand. Also, the squaring loop requires minimal hardware for cost reasons. A block diagram of a squaring loop is shown in Figure 5-7.

192   Transceiver System Design

The squaring loop squares the incoming signal, and filters the squared signal using a narrowband filter. The filter needs to be narrow enough for the spectrum of the input signal to be essentially constant. A matched filter gives the best signal/noise and should be used in the IF section before the squarer. This filter also reduces interference from other out of band signals and interference. The shape of the optimum matched filter should be a rectangle in the time domain (sinc function in the frequency domain). The low-pass filters have rectangular impulse responses with a time period of T seconds where T is one BPSK symbol interval. An integrate and dump circuit operates as a good matched filter. A two-pole Butterworth filter approximates the ideal matched filter. Squaring a direct sequence eliminates the modulation since the modulated signal is

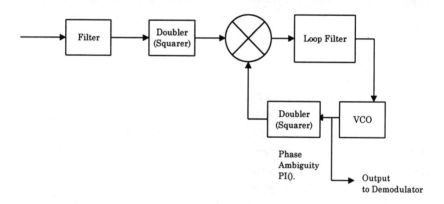

Figure 5-7 *Squaring loop carrier recovery.*

Demodulation

phase shifted either 0 or 180 degrees and squaring the signal doubles both the frequency and phase which results in the same phase shift of the signal as shown below:

$$S = A\cos(\omega t+\phi) \quad \phi = 0,180 \qquad 5.1$$

$$S^2 = A^2/2[1+\cos(2\omega t+2\phi)] \quad 2\phi = 0,360 \qquad 5.2$$

Therefore:

$$S^2 = A^2/2+A^2/2[\cos 2\omega t] \qquad 5.3$$

Therefore, squaring the signal gives rise to a DC term and a tone at twice the carrier frequency. A possible disadvantage to the squaring loop is that it needs to operate at twice the frequency. For most cases this is not a problem, however, higher frequencies can drive up the cost of hardware.

The output of the squaring loop is processed by a PLL and the output frequency is divided by two to achieve the fundamental frequency. However, the divider creates an ambiguity since a change in one cycle (2π radians) results in a change of phase of π. This means that if the frequency shifts one cycle, the phase is shifted 180 degrees. Since the PLL does not know when the cycle slips to another, the frequency is the same, therefore there is an ambiguity of $\pm f_c$. As a result, the phase of the demodulated signal can be $\pm$ the phase. This creates a possible problem in the data because of this phase reversal.

The output of squaring loop is phase-shifted to obtain a cosine wave, since the output of the PLL generates a sine wave, and is multiplied with the incoming signal to eliminate the carrier and obtain the data. If the signal is 90 degrees phase shifted with respect to the incoming signal, then the output would be zero:

$$(A\sin\omega t)(d(t)A\cos\omega t) = A/2 d(t)(\sin 0 + \sin 2\omega t)$$
$$= A/2 d(t)\sin 0 \quad \text{since } \sin 2\omega t \text{ is filtered}$$
$$= 0 \qquad\qquad 5.4$$

The phase shift is critical, may drift with temperature, and will need to be adjusted if different data rates are used. This is one of the main drawbacks to the squaring loop method of carrier recovery. There are other ways to perform carrier recovery that does not have this concern. The phase in question is the phase relationship between the input signal and the recovered carrier at the mixer where the carrier is stripped off.

Another disadvantage to the squaring loop approach to carrier recovery is that the matched filter for the squaring loop at the IF frequency is good for only one data rate. If the data rate is changed, the matched filter needs to be changed. Some systems require that the data rate is variable which makes the squaring loop less versatile than some other methods. A variable filter can be designed, but becomes more complex than a fixed filter.

If higher order PSK modulation schemes are used, then higher order detection schemes need to be implemented. For example, if a QPSK modulation waveform is used, then

Demodulation

the demodulation process needs to square the signal twice, or raise the signal to the fourth power. Sometimes this circuit is referred to as a times 4 device, which is really a misnomer. The signal is not times by 4, but the frequency is 4 times higher. The description of the squaring loop can be applied for all the higher order loops keeping in mind that the functions need to be modified for the higher order operations. The difference is how many times the frequency needs to be multiplied. For a n-phase signal, a n-times multiplier is required. The higher order PSK demodulation processes can become complex and expensive due to the higher frequency components that need to be handled. Also, there are losses in the signal amplitude everytime it is squared. These losses are called squaring losses and degrade the ability to detect the signal.

5.7.2 Costas Loop

The Costas loop does not use a squaring loop so the double-frequency signal is not a part of the process. Therefore, the higher frequencies are not produced. Instead, the Costas loop uses basically two phase lock loops, one of which is phase shifted in quadrature producing quadrature I and Q channels as shown in Figure 5-8.

The outputs of each phase lock loop is multiplied and low pass filtered (to eliminate the $f(t)^2$ terms) so the loop tracks sin(2P) where P is the phase error. The low pass filters are matched filters for the symbol or bit rate. To eliminate the need for the analog multiplier, a hard-limited Costas

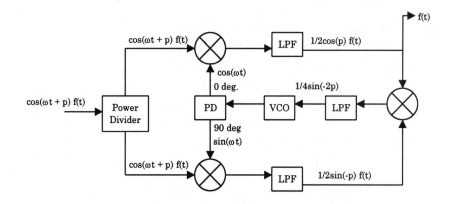

As p approaches zero the loop is locked. Note that the input could have been a sin instead of a cos and the bottom leg would have contained f(t).

Figure 5-8 *Costas loop used for carrier recovery.*

loop can be utilized by hard limiting the in-phase PLL path and then switching the sign of the error of the quadrature PLL. This becomes much easier to implement in hardware. This type of loop is commonly called a polarity loop. The matched filtering is done at baseband and the output is used to generate the error.

The phase shift, as in the squaring loop, is not critical since the feedback forces the phase to be correct. Since the feedback is forcing the error to be zero, the error channel is driven to zero but either channel could be the data channel. By hard limiting one of the channels, this automatically becomes the data channel since it just produces a sign change and the other channel becomes the error channel which is driven to zero.

Demodulation

The matched filters in the Costas loop are good for only one data rate. If the data rate is changed, then the matched filters need to be change. However, changing filters at baseband is much easier and more cost effective than changing the IF filters as in the squaring loop. Note that the squaring loop baseband matched filters also need to be changed along with the IF filter.

5.7.3 Modified Costas Loop and AFC Addition

Since the Costas loop is only good for a narrow bandwidth, an automatic frequency control (AFC) can be included to extend the pull-in range which increases the bandwidth of

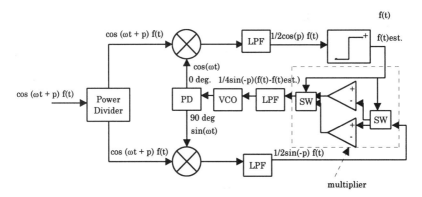

Figure 5-9 Hard-limited Costas loop for carrier recovery.

the Costas loop. There are other modifications of the standard Costas loop that can be done to improve the performance and versatility of this process. A hard-limited Costas loop or a data-aided loop can further improve the carrier recovery process. A hard-limited Costas loop has a hard limiter in one of the channels, that estimates the data pulse stream. This way the multiplier, which is generally a drawback to Costas loops, can simply invert or non-invert the signal (see Figure 5-9). This can be done by selecting either a non-inverting amplifier or an inverting amplifier according to the data estimate. This strips the modulation off and leaves the phase error for the PLL. A data-aided loop uses the data estimation similar to the hard-limited Costas to improve the performance of the standard Costas loop. There are many enhancements and variations of the standard Costas loop but the basic understanding of its operation provides the user the ability to design a carrier recovery loop for a typical system including modifications where improvement is desired.

5.8 Symbol Synchronizer

Once the carrier is eliminated, the raw data remains. However, due to noise and intersymbol interference (ISI) distorting the data stream, a symbol synchronizer is needed to determine what bit was sent. This device aligns the sample clock with the data stream so that it samples in the center of each bit in the data stream. An early-late gate can be used which is similar to the early-late gate used in the code demodulation process. The code demodulation process requires integration over the code

Demodulation   199

length or repetition to generate the autocorrelation function. With the bit synchronizer, the integration process is over a symbol or bit if coding is not used. The early and late gate streams in the bit synchronizer are a clock of ones and zeros since there is no code reference and the data is unknown. Therefore, the incoming data stream is multiplied by the bit clock, set at the bit rate with the transitions early or later than the data transitions as shown in Figure 5-10.

This example shows the early gate and late gates are off from the aligned signal by 1/4 symbol or bit. The integrated outputs are shown along with the aligned integration for comparison. When the integrated outputs are equal in peak amplitude as shown in the example, then the bit synchronizer is aligned with the bit. Therefore, the point in between the early gate and late gate, which is the center of the aligned pulse, is the optimal point to sample and recover the data. This point provides the best S/N of the received data. Once this data is sampled at the optimal place, a decision is made to determine whether a "1" or a "0" was sent. This bit stream of measured data is decoded to produce the desired data.

5.9   The Eye Pattern

The eye pattern is a description of the received digital data stream when observed on an oscilloscope. Due to the bandwidth limitations of the receiver, the received bit stream is filtered and the transitions are smoothed. Since

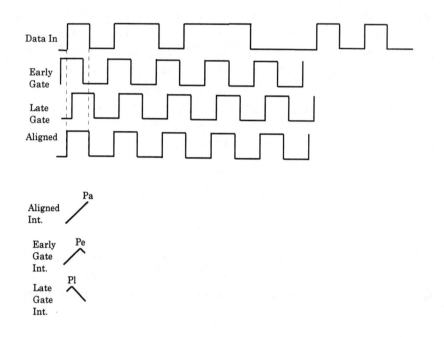

Figure 5-10 *Bit synchronizer using the early-late gate technique.*

the data stream is pseudo-random, the oscilloscope shows both positive and negative transitions, thus forming a waveform that resembles an eye (see Figure 5-11).

The four possible transitions at the corner of the eye are; low to high, high to low, high to high, and low to low. Observation of the eye pattern can provide a means of determining the performance of a receiver. The noise on the eye pattern and the closing of the eye can indicate that

Demodulation

the receiver's performance needs to be improved or that the signal from the transmitter needs to be increased. The eye pattern starts to close with the amount of distortion on the signal, see Figure 5-11. The eye pattern is what the bit synchronizer tries to sample in the center of the eye where the largest amplitude and the highest S/N occur.

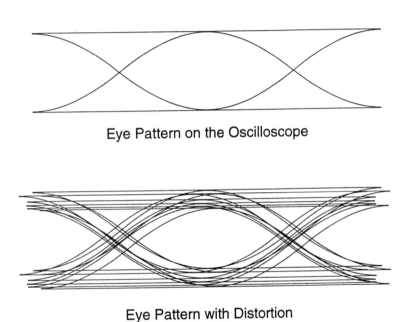

Eye Pattern on the Oscilloscope

Eye Pattern with Distortion

Figure 5-11 *Eye pattern as seen on an oscilloscope.*

5.10 Digital Processor

Once the digital data stream has been sampled and determined that a "0" or a "1" was sent, this digital data stream is decoded into the data message that was sent. The digital processor is generally responsible for this task and also the task of controlling the processes in the receiver and often the demodulation of the received signal. In many systems today, the digital processor, in conjunction with specialized digital signal processing (DSP) integrated circuits that are available, is playing more of a key role in the demodulation process. This includes implementation of the various tracking loops, bit synchronizing, carrier recovery, and decoding the data as mentioned above. The digital processor ensures that all of the functions occur at the necessary times by providing a control and scheduling capability for the receiver processes.

5.11 Intersymbol Interference

The intersymbol interference (ISI) is the amount of interference due to dispersion of the pulses to the other pulses in the stream. Dispersion occurs when there is non-linear phase responses to different frequencies, or a non-constant group delay which is the derivative of the phase. Since pulses are made up of multiple frequencies according to the Fourier series expansion of the waveform, these frequencies need to have the same delay through the system in order to preserve the pulse waveform. Therefore,

Demodulation

if the system has a constant group delay, or the same delay for all frequencies, then the pulse is preserved. If there is a non-constant group delay causing different delays for different frequencies, then when these frequencies are added together to form the pulse waveform where the pulse is dispersed or spread out. This causes distortion to adjacent pulses and therefore creates ISI. To obtain a value of the

amount of ISI, the eye pattern as mention above can be used. The ISI is determined by the following:

$$ISI = 20\log(V_h/V_l) \qquad 5.5$$

where:

V_h = highest peak voltage when measured at the center of the eye of the eye pattern.
V_l = lowest peak voltage when measured at the center of the eye of the eye pattern.

The measured amount of the ISI in a receiver determines the amount of increase that is required of E_b/N_o or S/N in the link budget described in Chapter 1. Sometimes this ISI is specified in the link budget as a separate entry.

5.12 Phase Shift Detection

Spread spectrum systems are sometimes used for covert systems so that the transmissions cannot be detected. This is known as electronic counter measures (ECM) and provides decreased vulnerability to detection. Receivers that are designed to detect these types of signals are called electronic counter counter measures (ECCM) receivers. One of the methods to detect a BPSK waveform as mentioned before is to use a squaring, doubler or × 2 (doubling the frequency) to eliminate the phase shift as follows:

$$A cos(\omega t + 0°, 180°)^2$$
$$= \frac{A^2}{2}(cos2\omega t + 2(0°, 180°)) + \frac{A^2}{2}(cos 0°)$$
$$= \frac{A^2}{2}(cos2\omega t + 0°) \quad \textit{plus d/c offset}$$

5.6

Note: 2×0 degrees is 0 degrees and 2×180 degrees is 360 degrees which is equal to 0 degrees. This is the basis for squaring, to eliminate the phase shift modulation so that the resultant signal is a CW spectral line instead of a sinc function with a suppressed carrier. The CW frequency is then easily detected since the despreads the wideband signal.

The end result is a spectral line at twice the carrier frequency. This is the basic principle behind the ECCM

Demodulation

receiver in detecting a BPSK data stream. This allows the ECCM receiver to know the frequency of the signal being sent by dividing the output frequency by two for BPSK. If BPSK is used, the bandwidth provides the chip rate that was sent if the sidelobes for BPSK are correct. Some filtering schemes alter both the bandwidths and the sidelobe levels.

QPSK and OQPSK are detected by what is known as a times 4 (×4) detector which basically quadruples the input signal to eliminate the phase ambiguities. Offset QPSK has all the same absolute phase states but is not allowed to switch more than 90 degrees which eliminates the 180 degree phase shifts. However, the criteria depends on only the absolute phase states and how they are eliminated. The following shows the results of quadrupling the signal:

$$A\cos(\omega t \pm 90°, +0°, 180°)^2$$
$$= \frac{A^2}{2}(\cos(2\omega t + 0°, 180,°)) \text{ plus d/c offset filtered out}$$

and squaring again $= \frac{A^4}{8}(\cos 4\omega t + 2(0°, 180°))$

plus d/c term filtered out

$$= \frac{A^4}{8}(\cos 4\omega t + 0°)$$

5.7

Note: Since the possible phase states are 0, 180, 90, −90, squaring them would only give 2×0 = 0, 2×180 = 360 = 0, 2×90 = 180, 2×−90 = −180 = 180. Therefore, the problem has been reduced to simple phase shifts of a BPSK level. One more squaring will result in the same as described in the BPSK example above which eliminates the phase shift. Therefore, quadrupling the signal eliminates the phase shift for a quadrature phase shifted signal.

The resultant signal gives a spectral line at four times the carrier frequency. This is the basic principle behind the ECCM receiver in detecting a QPSK or OQPSK data stream. This allows the ECCM receiver to know the frequency of the signal being sent by dividing the output frequency by four for both QPSK and OQPSK waveforms.

For MSK, there is a sinusoidal modulating frequency proportional to the chip rate in addition to the carrier frequency along with the phase transitions. One way to generate classical MSK is to use two BPSK in a OQPSK type system and sinusoidally modulating the quadrature channels at a frequency proportional to the bit rate before summation. In other words, the phase transitions are smoothed out by sinusoidal weighting. By putting this into a ×4 detector as above, the resultant is:

$$[A\cos(\omega t \pm 90°,+0°,180°)(B\cos\frac{\pi t}{2T})]^4 \qquad 5.8$$
$$=[\frac{AB}{2}(\cos(\omega \pm \frac{\pi}{2T})t+(\pm 90°,+0°,180°))]^4$$

From the QPSK example, the phase ambiguities will be eliminated due to the above equations since the signal is

Demodulation

quadrupled so that the phase is ×4 which results in 0 degrees, so these terms are not carried out. Squaring the equation first produces the sum and differences of the frequencies. Assuming that the carrier frequency is much larger than the modulating frequency, only the sum terms are consider and the modulating terms are filtered out. For simplicity, the amplitude coefficients are left out. Therefore squaring the above with the conditions stated results in:

$$\cos(2\omega \pm \frac{\pi}{T})t + \cos(2\omega)t \qquad 5.9$$

Squaring the above equation with the same assumptions will be the result of quadrupling MSK:

$$\cos 4\omega t + \cos(4\omega \pm 2\frac{\pi}{T})t + \cos(4\omega \pm \frac{\pi}{T})t \qquad 5.10$$

There will be spectral lines at 4 times the carrier, 4 times the carrier ± 4 times the modulating frequency, and 4 times the carrier ± 2 times the modulating frequency. Therefore, by quadrupling the MSK signal the carrier can be detected. Also, by the equation above, careful detection can produce the chip rate features since the modulating frequency is proportional to the chip rate.

5.13 Summary

The demodulation process is an important aspect in the design of the transceiver. Proper design of the demodulation section can enhance the sensitivity and performance of detection of the data. Two types of demodulation processes can be used to despread and recover the data. The matched filter approach simply delays and correlates each delay segment of the signal to produce the demodulated output. This process includes the use of PPM to encode and decode the actual data. Another demodulation process uses a coherent sliding correlator for despreading of the data. This process requires alignment of the codes in the receiver which is generally accomplished by a short acquisition code. Tracking loops such as the early/late gate aligns the code for the despreading process. Carrier recovery loops such as the squaring loop, Costas loop, etc., are required to provide a means for the demodulator to strip off the carrier. The symbol synchronizer is required to sample the data at the proper time in the eye pattern in order to minimize the effects of ISI. Finally, receivers designed for intercepting transmissions of other transmitters use various means of detection depending on the type of phase modulation utilized.

5.14 References

[1] Simon Haykin, *Communication Systems*, John Wiley & Sons Inc., New York, 1983.

[2] Jack K. Holmes, *Coherent Spread Spectrum Systems*, New York: Wiley & Sons, pp.251-267, 1982.

[3] M.K. Simon, J.C. Springett, "The Theory, Design, and Operation of the Suppressed Carrier Data-Aided Tracking Receiver", Technical Report 32-1583, Jet Propulsion Laboratories, 1973.

[4] Christopher R. Keate, "The Development and Simulation of a Single and Variable Data-Rate Data-Aided Tracking Receiver", Thesis Dissertation, 1986.

Problems

1. Given a code of 1010011, what would the weights of the pulse matched filter need to be assuming ×1 is the weight for the first stage of the matched filter and ×7 is the weight for the last stage of the matched filter?

2. Given a BPSK signal, show how a squaring loop eliminates the phase ambiguity.

3. What is the null-to-null bandwidth for a 50 Mcps chip rate?

4. How does the bandwidth change in (a) the pulsed matched filter and (b) the sliding correlator matched filter?

5. Explain what is meant by intersymbol interference.

6. Where on the eye pattern is the best place for sampling the signal for best performance? Where is the worst place?

7. Using the squaring loop idea for BPSK detection, what would be needed in an (a) 8-PSK detector, (b) 16-PSK detector?

8. What would be the minimum possible phase shifts of each of the waveforms in problem 6?

9. Since MSK can be considered as a frequency shifted signal, what would be another way of detection of MSK?

10. What is the advantage of MSK in reference to sidelobes?

6

Basic Probability and Pulse Theory

In order to achieve a better understanding of digital communications, some basic principles need to be discussed. This chapter gives an overview of theory necessary for the design of digital communication systems. This inlcudes a simple probability theory analysis, pulse analysis in both the time and frequency domain, and probability of error.

6.1 Simple Approach to Understanding Probability

Understanding probability is a key to designing digital communication systems and spread spectrum systems. Probability is used in the link budget calculations in regards to the error and required S/N ratio. There is even a probability of whether a transceiver is going to work and at what distances. To grasp the concept of probability and to get an intuitive feel for the probability process, a simple approach is provided below.

First of all, the question arises, what is the probability something is going to occur, whether it be an error in the system or the probability that multipath will prevent the signal from arriving at the receiver. The probability that an event occurs is called the Probability Density Function (PDF). The PDF is defined as:

$$f_X(x) \tag{6.1}$$

This can be looked at like a percentage, say 10 percent chance that a value is present or that an event has occurred. The integral of the density function equals 1, or in other words, the entire density function adds up to 100 percent.

$$\int f_X(X)dx = 1 \tag{6.2}$$

For example, if there is a 10 percent chance of getting it right, by default there is a 90 percent chance of getting it wrong. The PDF is the curve showing the probability of occurrence and is shown in Figure 6-1. Often times this curve is mislabeled as the distribution curve instead of the density curve. This is not the cumulative distribution curve, which will be discussed later in this chapter, but the PDF curve.

Basic Probability and Pulse Theory

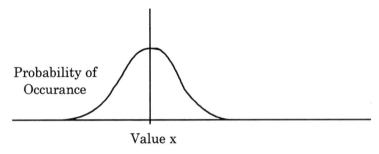

Probability Density Function for Gaussian Distribution

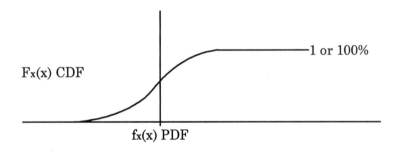

Cumulative Distribution Function for Gaussian Distribution

Figure 6-1 *Probability density function and cumulative distribution function for a Gaussian process.*

Another term used in probability theory is called the expected value. The expected value is the best guess as to what the value will be. To obtain the expected value, the signal value is multiplied by the percentage or the PDF and then sum them up or integrate the value as shown:

$$E[X] = \int x f_X(x) dx \qquad 6.3$$

Using a discrete example, suppose the density function is symmetrical. To plug in some numbers:

x	$f_x(x)$	$xf_x(x)$
1	.1(10%)	.1
2	.2(20%)	.4
3	.4(40%)	1.2
4	.2(20%)	.8
5	.1(10%)	.5

$\Sigma\, xf_x(x) = E[X] = 3.0$

The expected value is equal to the mean value or mean (m_x).

Therefore:

$$m_x = E[x] = \int x f_x(x) dx \qquad 6.4$$

Mean (m_x) = E[X] = 3.0

Basic Probability and Pulse Theory

Another important variable in probability theory is called the variance. In order to define the variance, the expected value of the variable squared needs to be calculated. The expected value of x^2 is solved by a similar method below:

$$E[x^2] = \int x^2 f_x(x) dx \qquad 6.5$$

Using the same discrete example:

x^2	$f_x(x)$	$x^2 f_x(x)$
1	.1	.1
4	.2	.8
9	.4	3.6
16	.2	3.2
25	.1	2.5

$\sum x^2 * f_x(x) = E[x^2] = 10.2$

The variance (Var) is a measure of how far the signal varies from the expected value or the mean. The variance is equal to:

$$E[x^2] - m_x^2 = 10.2 - 9 = 1.2. \qquad 6.6$$

If the mean is zero, then the variance is equal to:

$E[x^2] = 10.2.$

The variance is the expected value of the difference between the mean and the variable squared.

$$Var = \sigma_x^2 = E[(x-m_x)^2] = \int (x-m_x)^2 f_x(x)dx \qquad 6.7$$

For the example above:

$(x - mean)^2$	$f_x(x)$	$(x - mean)^2 f_x(x)$
4	.1	.4
1	.2	.2
0	.4	0
1	.2	.2
4	.1	.4

$$\Sigma (x - mean)^2 * f_x(x) = \quad 1.2$$

The standard deviation is a common term that also relates to the variation of the signal. The standard deviation is defined as:

$$StdDev = \sqrt{Var} = \sqrt{1.2} = 1.095 \qquad 6.8$$

The cumulative distribution function, or distribution function, accumulates or adds up the percentages between x1 and x2. It is defined as:

Basic Probability and Pulse Theory   219

$$F_x(x) = \int_{x1}^{x2} f_x(x)dx \qquad 6.9$$

For example, in the discrete case above, summing x1 = 1 to x2 = 3:

$$F_x(x) = .1+.2+.4 = .7 = 70\%. \qquad 6.10$$

The cumulative distribution function for a Gaussian process integrating over the entire range from $-\infty$ to $+\infty$ is graphically shown in Figure 6-1.

Note that the final value of the cumulative distribution function is equal to unity as explained previously. The distribution function is useful when the limits of integration are something other than $-\infty$ to $+\infty$. Finding out what the total probability of a range of signals can help to identify and analyze the strengths of a system. For example, often times an analysis requires to know what is the total probability would be over 2 σ, where σ is the standard deviation of the system. This requires calculation of the cumulative distribution function which will be discussed later. The power spectral density (PSD) is useful in communications systems to analyze the frequency domain of a signal. The PSD for a random binary wave BPSK is:

$$S_X(f) = A^2 T \text{sinc}^2(fT) \qquad 6.11$$

Therefore, the spectrum will be a sinc² function which is a (sinx/x)² function in the frequency domain.

The mean-square value $E[X^2(t)]$ is equal to the total area under the graph of the power spectral density (PSD):

$$E[x^2(t)] = \int S_x(f) df \qquad 6.12$$

If the mean is equal to zero, then this becomes the variance of the BPSK signal. The standard deviation is the square root of the variance.

Probability theory plays an important role in communications and further study should be done to enhance the ability to analyze digital transmissions. However, this simple presentation will greatly help in providing the understanding required to begin analyzing digital communications.

6.2 The Gaussian Process

The Gaussian process or distribution is probably the most common distributions in analyzing digital communications. There are many other types of distributions such as the

Basic Probability and Pulse Theory

Uniform distribution which is used for equal probability situations such as the phase of a multipath signal and the Rayleigh distribution which is used to characterize the amplitude of the multipath. However, the Gaussian distribution is used most often and is called the normal distribution or bell-shaped distribution because its common use and because it produces a PDF bell-shaped curve. Most often noise is characterized using a Gaussian distribution for transceiver performance and S/N evaluations. The probability density function for a Gaussian process is defined as shown below:

$$f_x(x) = \frac{1}{\sqrt{2\pi}\sigma} e^{\frac{-(x-m_x)^2}{2\sigma^2}}$$ 6.13

This establishes the curve that is called the distribution function. This is really the density function, a slight misnomer. Note that it is not the cumulative distribution function as mentioned before. The cumulative distribution is defined as:

$$F_x(x) = \frac{1}{2}[1 + erf\frac{x}{\sqrt{2}\sigma}]$$ 6.14

where:

mean = 0

This value can be calculated or a lookup table is used with an approximation if x is large. The cumulative distribution function is used to calculate the percent that the error is within a given range, such as 1-sigma or 2-sigma value.

For example, what is the probability that x is between ± 2 sigma or a 2 sigma variation or 2 times the standard deviation. The cumulative distribution is used and substituting −2 sigma for x in the above equation gives:

$$F_x(x) = \frac{1}{2}[1+erf-\frac{2}{\sqrt{2}}] = \frac{1}{2}[1+erf-\sqrt{2}] \qquad 6.15$$

Note that erf(−x) = −erf(x). Therefore, erf(−2^(1/2)) = −erf(1.414) = −.954 from tables, see Appendix 2 *Communication Systems* (Haykin). Finishing the calculations:

$$F_x(x) = \frac{1}{2}[1 + (-.954)] = .023. \qquad 6.16$$

This represents the probability of the function being at −2 σ or less as shown in Figure 6-2. Therefore, assuming gaussian distribution with zero mean, the probability of the function being a +2 σ or greater is the same.

Therefore, the probability of the signal being outside ± 2 sigma is 0.0456. Therefore, the probability of x being inside these limits is 1−.0456 = 0.954 or 95.4%. This was the same result as simply taking the erf function above,

Basic Probability and Pulse Theory 223

erf$\sqrt{2}$ = 0.954. For a −1-sigma error, the error function of −0.707 is −0.682 from tables and the distribution function is 0.159 as shown in Figure 6-2.

Therefore, for ± 1 sigma, the probability of being outside these limits is 0.318. The probability of being within these limits is 1−.318 = 0.682 = 68.2%. Therefore, given the range of x, the probability that the solution falls within the range is easily calculated.

One caution on using the Gaussian distribution is that if the number of samples is not infinite, then the Gaussian limit may not be accurate and will degrade according to the number of samples that are taken.

6.3 Quantization Error

Since all digital systems are discrete and not continuous, the digitizing process creates errors between the digital samples. This error is created because a digitizing circuit

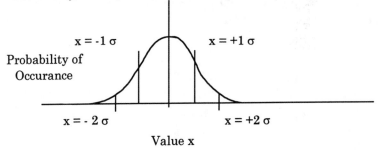

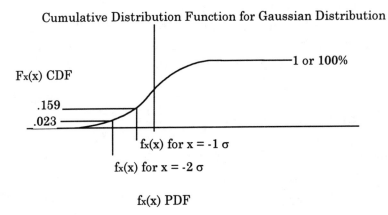

Figure 6-2 *Probability within a given range using Gaussian distribution.*

is trying to simulate a continuous signal or represents time which is continuous. If a signal is changing continuously with time, then there is an error on the estimation of the signal between the sample points. Obviously, if the sample rate is increased, this error is reduced.

Basic Probability and Pulse Theory

For example, range is based on time of arrival (TOA) of the pulse that is sent and returned. This is the basic principle that ranging and tracking radars work on when calculating the range of an aircraft from the ground station. One source of range error is called quantization error. If the arrival pulse is quantized or sampled, the quantization error is based on the distance between sample pulses, half way on either side is worse case quantization error as shown in Figure 6-3.

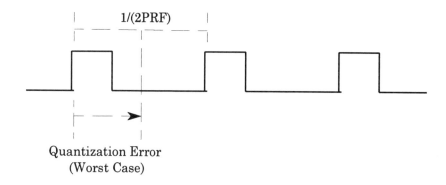

Figure 6-3 *Quantization error showing worst case conditions.*

Since the probability that this occurrence is equal across the worse case points, then it is said to be uniform or have a uniform distribution. The variance for a uniform distribution is:

$$\sigma^2 = \frac{a^2}{3} \qquad 6.17$$

where distribution varies between ± a.

The standard deviation is equal to the square root of the variance and is shown by the following:

$$\sigma = \sqrt{\frac{a^2}{3}} \qquad 6.18$$

For example, if the clock rate is 50 MHz, the time between pulses is 20 ns so "a", or peak deviation, is equal to 10 ns. Therefore the standard deviation is equal to:

$$\sigma = \sqrt{\frac{10^2}{3}} = 5.77 ns \qquad 6.19$$

If a number of samples are taken and averaged out, (integrated), then the standard deviation is reduced by:

Basic Probability and Pulse Theory

$$\sigma_{ave} = \frac{\sigma}{\sqrt{n_{samples}}} \qquad 6.20$$

If the number of samples that are to be average is 9, then the standard deviation is:

$$\sigma_{ave} = \frac{5.77 ns}{\sqrt{9}} = 1.92 ns \qquad 6.21$$

In order to determine the overall error, the distribution from each source of range error needs to be combined. This combination is accomplished by doing a root sum squared (RSS) solution. This is done by squaring each of the standard deviations, summing, and then taking the square root of the result for a final error. For example, if one error has a $\sigma = 1.92$ and another independent error has a $\sigma = 1.45$, then the overall error is:

$$\text{Total Error } \sigma = \sqrt{1.92^2 + 1.45^2} = 2.41 \qquad 6.22$$

This combination of the uniform distributions result in a normal or Gaussian distribution overall. Therefore, even though each error is uniform and is analyzed using the uniform distribution, the resultant error is Gaussian and follows the Gaussian distribution for analysis. However, this assumes independent sources of error. In other words, one source of error cannot be related or dependent on another source of error.

6.4 Probability of Error

The performance of the demodulation is measured in either BER (bit error rate), or POE (probability of error) as shown:

$$BER = Err/TNB \qquad 6.23$$

$$POE = 1/2 \; erfc \; (E_b/N_o)^{1/2} \qquad 6.24$$

where:

Err = number of bit errors
TNB = number of total bits
POE = probability of error
E_b = energy in a bit
N_o = noise power spectral density (noise power in a 1 Hz bandwidth)
$erfc$ = complimentary error function

The BER is simply calculated by adding up the number of bits that were in error and dividing by the total number of bits for that particular measurement. BER counters continuously calculate this ratio. The bit error rate could have been called a bit error ratio since it is more related to the ratio than it is to time.

Normally, the probability of error P_e (POE) is calculated from the measured E_b/N_o for a given type modulation. The

Basic Probability and Pulse Theory

probability of getting a zero when actually a one was sent for a BPSK modulated signal is the $P_{e1}=1/2\,\text{erfc}((E_b/N_o)^{1/2})$. The probability of getting a one when a zero was sent is $P_{eo}=1/2\,\text{erfc}((E_b/N_o)^{1/2})$. The average of these two probabilities is $P_e=1/2\,\text{erfc}((E_b/N_o)^{1/2})$. Note that the average probability is the same as each of the individual probabilities. The reason for this is the probabilities are the same and are independent, that is, either a one was sent and interpreted as a zero, or a zero was sent and interpreted as a one. The probabilities happen at different times. P_e for other systems are as follows:

Coherent FSK: $P_e=1/2\,\text{erfc}((E_b/2N_o)^{1/2})$ 6.25

Noncoherent FSK: $P_e=1/2\,e(-E_b/N_o)$ 6.26

DPSK (noncoherent): $P_e=1/2\,e(-E_b/N_o)$ 6.27

Coh. QPSK: $P_e=1/2\,\text{erfc}((E_b/N_o)^{1/2})-1/4\,\text{erfc}^2((E_b/N_o)^{1/2})$ 6.28
Note: The second term can be eliminated for $E_b/N_o \gg 1$.

Coherent MSK: Same as Coherent QPSK.

Plotting these curves using a spreadsheet is shown in Figure 6-4.

Note that the coherent systems provide a better P_e for a given E_b/N_o, or less E_b/N_o is required for the same P_e. Also, note that coherent BPSK and coherent QPSK are fairly close at higher signal levels and diverge with small E_b/N_o ratios.

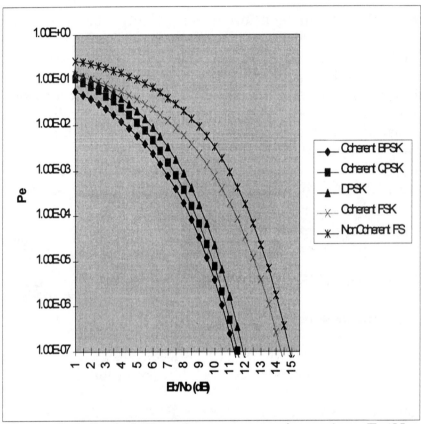

Figure 6-4 *Probability of error curves for a given E_b/N_o.*

Basic Probability and Pulse Theory

6.5 Probability of Detection and False Alarms

Some systems, such as radar and pulse position modulation systems, do not use just probability of error since data is not sent all the time. It is not a matter of just whether a zero or a one was sent, but whether anything was sent at all. Therefore, errors can come from missed detection or false detection.

The time between false alarms T_{fa} is dependent on the integration time which determines the number of pulses integrated N_{pi}. The number of pulses integrated is divided by the probability of false alarm P_{fa} as shown:

$$T_{fa} = N_{pi}/P_{fa} \qquad 6.29$$

Therefore, the longer the integration time, the less false alarms will be received in a given period of time.

For a pulsed system, there will be times when nothing is sent and the detection process needs to detect whether or not anything was sent. There is a threshold that is set in the system to detect the signal. If nothing is sent and the detection threshold is too low, then the probability of false alarms or detections will be high when there is no signal present. If the detection threshold is set too high to avoid false alarms, then the probability to detect the signal when it is present will be too low. If the probability of detection is increased to ensure that we detect the signal, then there is a more likely chance that we will detect noise which increases the probability of false alarm. There is a tradeoff that specifies where to set the threshold. Notice that the probability of detection is whether or not the signal was detected and probability of false alarm is whether or not the noise was detected. This should not be confused with whether a zero or a one was detected which is evaluated using probability of error. A curve showing the probability density functions of one sided noise and signal plus noise is shown in Figure 6-5.

The energy is selected to provided the desired probability of detection and probability of false alarm. The energy is the energy contained in the pulse E_p. The cumulative distribution functions are used to calculate these

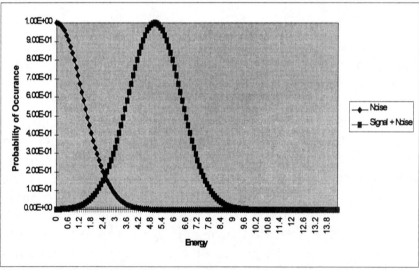

Figure 6-5 *Probability density functions of noise and signal plus noise.*

Basic Probability and Pulse Theory 233

parameters. For example, if the threshold is selected at 2.4 in Figure 6-5, then the cumulative distribution function from 2.4 to +∞ (area under the Signal + Noise curve from 2.4) is the probability of detection. The probability of false alarm is the cumulative distribution function from 2.4 to +∞ (area under the noise curve from 2.4).

The cumulative distribution function for Gaussian distribution is used since these processes are Gaussian.

Since the P_{fa} and P_d are integrating the function from 2.4 to +∞, and the total distribution function is equal to unity, then the distribution function is equal to:

$$F_x(x) = 1 - (\frac{1}{2}[1 + erf\frac{x}{\sqrt{2}\sigma}]) = 1/2(1 - erf\frac{x}{\sqrt{2}\sigma}) \qquad 6.30$$

The complementary error function is equal to:

$$erfc = 1 - erf \qquad 6.31$$

Therefore, by substituting the erfc in the equation above produces:

$$F_x(x) = 1/2(erfc\frac{x}{\sqrt{2}\sigma}) \qquad 6.32$$

The variance for both P_{fa} and P_d is:

$$\text{VAR} = E_p N_o/2 \qquad 6.33$$

The standard deviation or σ is the square root of the variance:

$$\text{STD DEV.} = (E_p N_o/2)^{1/2} \qquad 6.34$$

Substituting the standard deviation into the above equation gives:

$$F_x(x) = 1/2(erfc\frac{x}{\sqrt{E_p N_o}}) = P_{fa} \qquad 6.35$$

Since the P_d contains an offset of the actual standard deviation from zero, this offset is included in the equation as follows:

$$F_x(x) = 1/2(erfc\frac{x}{\sqrt{E_p N_o}} - \sqrt{E_p/N_o}) = P_d \qquad 6.36$$

This P_d is the probability of detecting one pulse at a time. If there are 10 pulses required for a system to receive, and each pulse is independent and have the same probability

Basic Probability and Pulse Theory

for detection, then the probability of detecting all 10 pulses is $(P_d)^{10}$ for the 10 independent probabilities for each pulse sent. Often times error correction is used to provide increased probability detection of each of the pulses. In many systems, the probability of missing a pulse needs to be determined and can be calculated by probability theory using the Binomial Distribution Function (BDF).

6.6 Pulsed System Probabilities using Binomial Distribution Function

The Binomial Distribution Function is used to calculate the probability of missing a pulse as mentioned above. For example, suppose that the probability of detection is 99% for one pulse according to the criteria for probability of detection as described in section 6.5. If a message consists of 10 pulses, then the probability of getting all 10 pulses is $.99^{10} = .904$ or 90.4%.

The probability of missing one pulse is determined by using the binomial theorem as follows:

$$p(9) = \binom{10}{9} p^9 (1-p)^{(10-9)} = \binom{10}{9}(.99)^9(.01)^1$$
$$= \frac{10!}{(10-9)!9!}(.99)^9(.01)^1 = 9.1\%$$

6.37

Therefore, the percentage of errors that result in only one pulse lost out of 10 is:

Percent(one pulse lost) = 9.1%/(100%–90.4%) = 95%.

Therefore, if there are errors in the system, 95% of the time it will be caused by one missing pulse. By the same analysis, the percent of the time that the errors in the system are caused by two pulses missing is 4.3% and so forth for each of the possibilities of missing pulses. The total possibilities of errors are summed to equal 100%.

6.7 Error Detection and Correction

Error detection and correction are used to improve the integrity and continuity of service for a system. The integrity is concerned with how reliable is the received information. That is, if information is received, how true is the data and is it accurate data and not false or noise data. Error detection can be used to inform the system that the data is false. Continuity of service is concerned with how often the data is missed which disrupts the continuity of the data stream. Error correction is used to fill in the missing or corrupted data to enhance continuity of service.

There are many techniques used to perform both the error detection and error correction. The concern in designing a transceiver is what are the requirements and how much overhead can the system tolerate. Note that the error detection and error correction require extra bits to be sent

Basic Probability and Pulse Theory

that are not data bits. Therefore, the amount of error detection and error correction can significantly reduce the data capacity of a system which is a tradeoff in the system design.

6.8 Sampling Theorem and Aliasing

Nyquist states that an analog function needs to be sampled at a rate of at least 2 times the highest frequency component in the analog function to recover the signal. If the analog function is sampled less than this, aliasing can occur.

When a signal is sampled, harmonics of the desired signal are produced. These harmonics produce replicas of the desired signal (positive and negative frequencies) in the frequency domain at $1/T_s$ intervals. When the analog signals are undersampled, the harmonics of the repeated signal start to move into the passband of the desired signal and produce aliasing. Pre-aliasing filters as mentioned in Chapter 3 can eliminate this problem for digital sampling and transmissions.

6.9 Theory of Pulse Systems

The easiest way to explain what is happening with a spread spectrum waveform is to first of all look at a simple square wave. A square wave has a 50% duty cycle and is shown in Figure 6-6.

This is the time domain representation of the square wave because it shows the amplitude of the signal with respect to time. This can be observed on a oscilloscope. The frequency domain representation of the square wave shows the amplitude (usually in dB) as a function of frequency as shown in Figure 6-6. The frequency spectrum of the square wave contains a fundamental frequency, both positive and negative, and the harmonics associated with the fundamental frequency. The negative frequencies do not exist, they are there to help in analyzing the signal when it is upconverted. The fundamental frequency is the frequency of the square wave and if the corners of the square wave were smoothed to form a sine wave, this would be the fundamental frequency in the time domain. The harmonics create the sharp corners of the square wave. The sum of all these frequencies are usually represented by a Fourier series. The Fourier series gives all the frequency components with their associated amplitudes.

In a square wave, even harmonics are all suppressed, and the amplitudes of the odd harmonics form a sinc function as shown in Figure 6-6. The inverse of the pulse width is where the null of the main lobe is located. For a square wave, the null is located right at the 2nd harmonic which is between the fundamental frequency and the 3rd harmonic. Note that all the other nulls occur at the even harmonics since the even harmonics are suppressed with a square wave (see Figure 6-6).

Basic Probability and Pulse Theory

Time Representation of a Square Wave

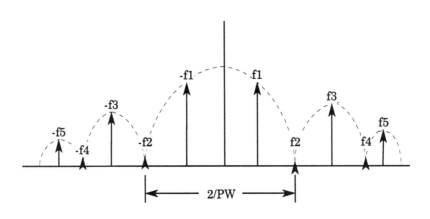

Frequency Spectrum of a Square Wave.

Figure 6-6 *Time and frequency domain representations of a pulsed signal with 50% duty cycle.*

If the pulse stream is not a square wave (50% duty cycle), then these frequency components get shifted around (see Figure 6-7).

If the pulse width is unchanged and the duty cycle is changed, then only the frequency components are shifted around and the sinc function is unchanged. Also, the frequency components that are suppressed are different with different amplitude levels. For example, if the duty cycle is changed to 25%, then every other even harmonic (4,8,12) is suppressed see Figure 6-7.

If the pulse width changes, then the sinc function envelope changes since the first null occurs at 1/(pulse width) as shown in Figure 6-8 for a 12.5% duty cycle.

6.10 PN Code

To get an intuitive feel for what the spectrum of a PN-coded signal would be, as the code is changing, the pulse widths change (variable number of 1's in a row) and the duty cycle is changing (variable number of −1's in a row). Therefore, the variation in the pulse widths causes the sinc function to be changing (will not be any larger than 1/chip width), and the variation in duty cycle causes the number of frequency components inside the sinc function will be changing. Therefore a more continuous sinc function will be the results with the null of the sinc function being 1/(chip width).

Basic Probability and Pulse Theory

Time Representation of a Pulse with 25% Duty Cycle

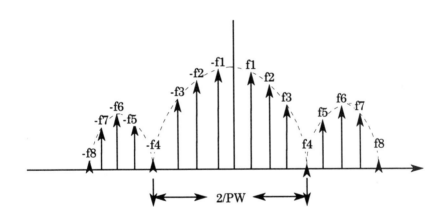

Frequency Spectrum of a Pulse with 25% Duty Cycle

Figure 6-7 *Time and frequency domain representations for 25% duty cycle pulse.*

Time Representation of a Pulse with 1/2 the width and
with a resultant duty cycle of 12.5%.

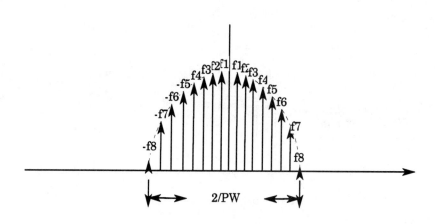

Frequency Spectrum of a Pulse with 12.5% Duty Cycle

Figure 6-8 *Time and frequency domain representation of 12.5% duty cycle pulse.*

Basic Probability and Pulse Theory 243

6.11 Summary

A simple approach to understanding probability theory and the gaussian process is provided to allow the designer to better understand the principles in designing and evaluating digital transmission. Also, quantization error is discussed since all digital systems need to include this error in the analysis. The probability of error and probability of detection and false alarm are the keys in determining the performance of the receiver. These errors are dependent on the received S/N and the type of modulation used. Errors are evaluated and error correction is used to mitigate the errors. Sampling theorem, aliasing, and theory on pulsed systems showing time and frequency domain plots provide the designer with knowledge and insight to design an optimum digital transceiver.

6.12 References

[1] Simon Haykin, *Communication Systems*, John Wiley & Sons Inc., 1983.

[2] Sheldon Ross, *A First Course in Probability*, Macmillan, New York, 1984.

Basic Probability and Pulse Theory

Problems

1. Using the fact that the integral of the Probability Density Function is equal to 1, what is the percent chance of getting this problem wrong if the chance of getting it right is 37%?

2. What is the expected value of x if $f_x(x) = .4$ when x $= 1$ and $f_x(x) = .6$ when x $= 2$? What is the mean?

3. Is the answer in problem 2 closer to 1 or 2 and why?

4. What is the $E[x^2]$ in problem 2?

5. What is the variance in problem 2?

6. What is the standard deviation of problem 2?

7. What is the probability the signal will not fall into a 2 σ value with a Gaussian distribution?

8. Given a system with too much quantization error, name a design change that can reduce the error by 50%?

9. What is the probability of receiving 20 pulses if the probability of detection is .98 for one pulse?

10. What is probability that the error occurs because of 1 lost pulse in problem 9 above?

11. What is the null-to-null bandwidth of a spread spectrum signal with a chipping rate of 50 Mcps?

7

Multipath

Multipath is defined as a free space signal transmission path that is different than the desired, or direct free space signal transmission path, in radar and communications applications. The amplitude, phase and angle of arrival of the multipath signal interferes with the amplitude, phase and angle of arrival of the desired, or direct path signal. This interference can create errors in angle of arrival information and in received signal amplitude in some radar applications. Angle of arrival errors are called "glint" errors. Amplitude fluctuations are called "scintillation" or "fading" errors. Therefore, the angle of arrival, the amplitude and the phase of the multipath signal are all critical parameters to consider when analyzing the effects of multipath signals in radar receivers.

Frequency diversity and spread spectrum systems contain a degree of immunity from multipath effects since these effects vary with frequency. For example, one frequency component for a given range and angle may have multipath that severely distorts the desired signal whereas another frequency may have little effect. This is mainly due to the difference in wavelength of the different frequencies.

7.1 Basic Types of Multipath

Multipath reflections can be separated into two types of reflections, specular and diffuse, and are generally a combination of both specular and diffuse. Specular multipath is a coherent reflection, which means that the phase of the reflected path is relatively constant with relation to the phase of the direct path signal. This type of reflection usually causes the greatest distortion of the direct path signal because most of the signal is reflected towards the receiver. Diffuse multipath reflections are noncoherent with respect to the direct path signal. The diffuse multipath causes less distortion than the specular type of multipath because it reflects less energy towards the receiver and usually has a noise-like response due to the random dispersion of the reflection. Both types of multipath can cause distortion to a system which increases the error of the received signal and reduces coverage by affecting the link budget. Specular reflection is analyzed for both a reflection off a smooth surface and a rough surface.

7.2 Specular Reflection on a Smooth Surface

Specular reflections actually occur over an area of the reflecting surface (which is defined as the first Fresnel zone, similar to the Fresnel zones found in optics). Generally the reflecting area is neglected and geometrical rays are used. These rays obey the laws of geometrical optics where the angle of incidence is equal to the angle of reflection, as shown in Figure 7-1.

Multipath

For a strictly specular reflection (referred to as a "smooth" reflection), the reflection coefficient ρ_o depends on the grazing angle, the properties of the reflecting surface, and the polarization of the incident radiation. This smooth reflection coefficient is complex. A smooth reflection has only one path directed towards the receiver as shown in Figure 7-1.

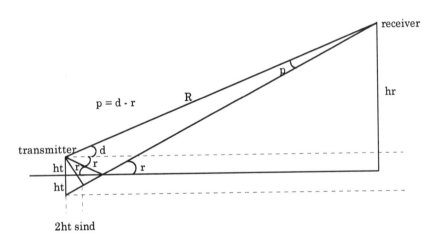

Specular analysis, angle of incident = angle of reflection.

$d = \arcsin[(hr - ht)/R]$

$r = \arcsin[(hr+ht)/(R+2ht \sin d)]$

Figure 7-1 *Multipath showing one single reflected ray.*

The magnitude of the smooth reflection coefficient is plotted in Figure 7-2 as a function of the grazing angle and polarization of the incident radiation.

The graph shows that for vertical polarization near the pseudo-Brewster angle the reflection coefficient is very small, a phenomenon that is also observed in optics. This phenomenon is used by some radars to minimize multipath reflections. For horizontal polarization, the reflection coefficient is fairly constant, but starts dropping off as the grazing angle approaches 90 degrees. Not only does the

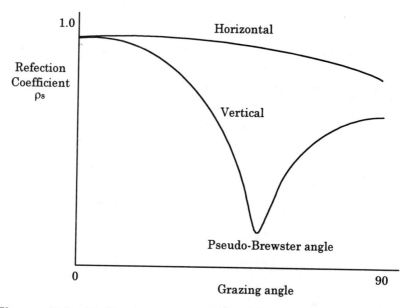

Figure 7-2 *Reflection coefficient vs. grazing angle for a smooth plane earth.*

Multipath

magnitude of the reflected radiation change, but the phase of the reflected radiation is also modified. A phase shift of the incident radiation can vary from near 0 degrees to 180 degrees, depending on the polarization of the incident radiation and the conductivity and dielectric constant of the reflecting surface.

7.3 Specular Reflection on a Rough Surface

Generally, the ideal reflecting case must be modified due to the fact that the reflecting surface (usually the Earth's surface) is not perfectly smooth. The roughness of the reflecting surface will decrease the amplitude of the reflection coefficient by scattering some energy in directions other than the direction of the receiving antenna. To account for the loss in received energy, a "scattering" coefficient, ρ_s, is used to modify the smooth reflection coefficient. The smooth reflection coefficient is multiplied by the scattering coefficient, creating an overall modified reflection coefficient to describe a specular surface reflection on a rough surface. The scattering coefficient, ρ_s, is the rms value. This coefficient can be defined as a power scattering coefficient as shown:

$$\overline{\rho_s^2} = e^{[-(\frac{4\pi h \sin d}{\lambda})]^2} \qquad 7.1$$

where:

h = rms height variation (normally distributed)

The roughness of the reflecting surface modifies the smooth reflection coefficient and produces a non-coherent reflection which is diffuse. If the variations are not too great, then a specular analysis can still be done and ρ_s will be the scattering modifier to the smooth specular coefficient ρ_o. This analysis can be used if the reflecting surface is smooth compared to a wavelength. The criteria that is used to determine if the diffuse multipath can be neglected, is known as the Rayleigh criteria, which looks at the ratio of the wavelength of the radiation and the height variation of the surface roughness and compares that to the grazing angle or incident angle. The Rayleigh criteria is as follows:

$$h_d \sin d < \frac{\lambda}{8} \qquad 7.2$$

where:
h_d = peak variation in the height of the surface.
d = grazing angle.

If the Rayleigh criteria is met, then the multipath is a specular reflection on a rough surface and ρ_s and ρ_o are used to determine the multipath.

If the Rayleigh criteria is not met, then the diffuse multipath coefficient ρ_d must be taken into account for complete multipath analysis.

7.4 Diffuse Reflection

Diffuse multipath is non-coherent reflections and is reflected from all or part of an area known as the "glistening surface" (see Figure 7-3).

This surface is called a glistening surface because in the optical (visible) equivalent the surface can be seen to sparkle or "glisten" when diffuse reflections are present.

The boundaries of the glistening surface are given by:

$$y = \pm \frac{X_1 X_2}{X_1 + X_2} \left(\frac{h_r}{X_1} + \frac{h_t}{X_2} \right) \sqrt{\beta_o^2 - \frac{1}{4}\left(\frac{h_r}{X_1} - \frac{h_t}{X_2}\right)^2} \qquad 7.3$$

Diffuse reflections are treated as having random amplitudes and phases with respect to the amplitude and phase of the direct path. The amplitude variations are Rayleigh distributed, while the phase variations are uniformly distributed from 0 degrees to 360 degrees. The phase distribution has zero mean, so that diffuse multipath appears as a type of noise to the receiving radar system.

For diffuse reflection, a diffuse scattering coefficient, ρ_d, is multiplied by ρ_o to give the diffuse reflection coefficient. Calculating the diffuse scattering coefficient ρ_d, is more tedious than calculating the specular scattering coefficient, ρ_s. The area that scatters the diffuse multipath cannot be

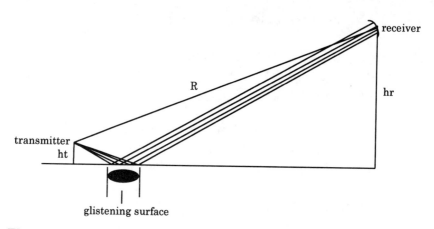

Figure 7-3 *Diffuse reflection is analyzed over a glistening surface.*

neglected as was done in the specular multipath case. One procedure to determine ρ_d is to break up the glistening area into little squares, each square a reflecting surface, and then calculating a small (delta) diffuse scattering coefficient [2]. The scattering coefficients are summed together and the mean is calculated, which gives the overall diffuse scattering coefficient, ρ_d. Further analysis converts this to a power coefficient as shown below:

$$\rho_d^2 \approx \frac{1}{4\pi\beta_o^2} \int \frac{R^2 dS}{X_1^2 X_2^2} = \frac{1}{2\pi\beta_o^2} \int \frac{R^2 y dX}{(R-X)^2 X^2} \qquad 7.4$$

This diffuse scattering coefficient is derived based on the assumption that the reflecting surface is sufficiently rough so that there is no specular multipath component present.

Multipath

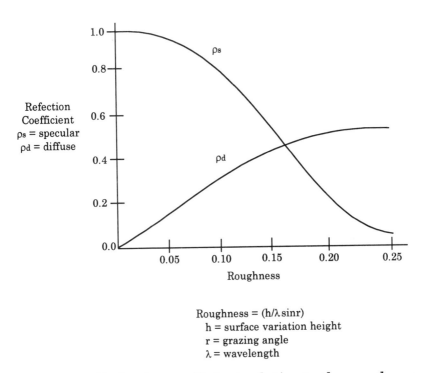

Figure 7-4 *Reflection coefficient relative to the roughness of the surface.*

The roughness factor has been defined as, F_d^2, which modifies the diffuse scattering coefficient, ρ_d. F_d^2 is equal to $1 - \rho_s^2$. A plot of the specular and diffuse scattering coefficients with respect to a roughness criteria is shown in Figure 7-4. With a roughness factor greater than .12, the reflections become predominantly diffuse. The diffuse reflection coefficient amplitude only reaches approximately 0.4 which is 40% of the amplitude of the smooth reflection coefficient whereas the specular coefficient reaches 1.0 or

100% of the smooth reflection coefficient. Precautions need to be taken when using the plots in Figure 7-4 since ρ_d does not include the roughness factor, F_d^2. This may cause significant errors in the analysis for actual values.

7.5 Curvature of the Earth

For most radar and communication systems other than satellite links, the divergence factor D caused by the curvature of the earth can be taken as unity. However, if a particular scenario requires the calculation of the diversion factor, it is equal to:

$$D = \lim_{f \to \infty} \sqrt{\frac{A_r}{A_f}}$$

A diagram showing the different areas that are projected for both round and flat earth are shown in Figure 7-5.

Basically, the curvature of the earth produces a wider area of reflection as shown in the figure. The ratios of the areas are used to calculate the divergence factor. Note that for most system the area difference is so small that it can be neglected.

Multipath

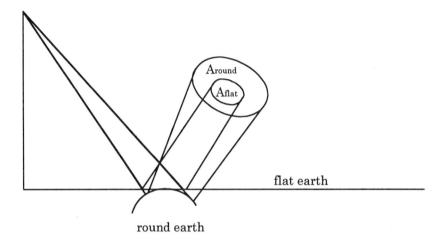

A_{round} = area projected due to a round earth.

A_{flat} = area projected due to a flat earth.

Figure 7-5 *Different areas for both the round and flat earth.*

7.6 Pulse Systems (RADAR)

Multipath can affect a radar system by interfering with the desired signal. Leading edge tracking can be used to eliminate a large portion of the distortion caused by multipath. This is because the multipath arrives later in time than the direct path. In leading edge tracking, only the first portion of the pulse is processed and the rest of the pulse, which is distorted due to multipath, is ignored. If the grazing angle is very small, then the time of arrival for both direct path and multipath signals are about the same. For this case the angle error between the direct path and multipath is very small. If the multipath is diffuse, then the multipath signal will appear as an increase in noise floor of the direct path and a slight angle variation will occur due to the multipath mean value coming at a slightly different angle. If the multipath is specular, then the multipath signal will increase or decrease the actual amplitude of the direct path signal and also create a slight angle error. If one or more of the antennas are moving, then the specular reflection will also be a changing variable with respect to the phase of the multipath signal and the direct path signal, thus producing a changing amplitude at the receiver. As the grazing angle increases, the angle error also increases. However, the time of arrival increases which means that leading edge tracking becomes more effective with larger grazing angles.

One other consideration is that the pulse repetition frequency (prf) of the radar be low enough so that the time delay of the multipath is shorter than the time between pulses transmitted. This prevents the multipath return from interfering with the next transmitted pulse.

Another consideration is the scenario of having a surface such as a building or other smooth surface giving rise to a specular reflection with a high reflection coefficient. The multipath could become fairly large in this case. For a

Multipath

stationary situation this is a real problem. However, if one of the antennas is moving, then the multipath is hindering for only the time the angle of reflection is right. This depends on the velocity of the antenna and the area of the reflector. The processor for the receiver could do an averaging of the signal over time and eliminate some of these problems.

7.7 Vector Analysis Approach

Different approaches can be taken to calculate the resultant effects of the reflected energy [3]. One way is to consider vector addition. The reference vector is the direct path vector with a given amplitude and a 0 reference angle. The coherent reflection vector, C, for specular reflection is calculated below with the phase referenced to the direct path vector:

$$\overline{C} = \overline{D} \overline{\rho_o} \overline{\rho_s} e^{j2\pi \frac{dR}{\lambda}}$$

7.5

Where:

D is the diversion factor
ρ_o is the specular reflection coefficient
ρ_s is the scattering coefficient
dR is the path length difference
λ is the wavelength

Determining a accurate vector for ρ_s is a difficult item. Note that ρ_s represents an average amplitude distributed value that is a function of the distribution of the surface.

The non-coherent vector, I, for diffuse reflection is calculated by:

$$\bar{I} = \frac{\overline{D\rho_o\rho_d}}{\sqrt{2}}(I_1 + jI_2)e^{j2\pi\frac{dR}{\lambda}} \qquad 7.6$$

The vectors C and I are summed with the reference phasor to produce the resultant phasor as shown in Figure 7-6.

Again, coming up with an accurate model for ρ_d is difficult. Observed test data from known systems can aid in the selection and verification of the scattering coefficients, ρ_s and ρ_d.

7.8 Power Summation Approach

Another approach in determining the error caused by reflections is to sum up the powers of each type of reflection with the direct power as shown:

$$P_t = P_{dir} + P_{spec} + P_{diff} \qquad 7.7$$

where:

P_{dir} = direct power

Multipath

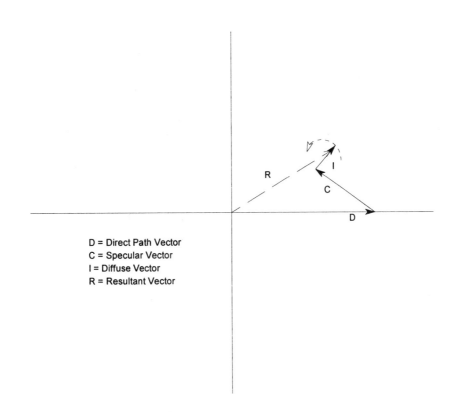

Figure 7-6 Phasor representation of multipath.

P_{spec} = mean specular power
P_{diff} = mean diffuse power

This approach uses power addition instead of vector addition. Mean power reflection coefficients are used. The resultant mean specular power is:

$$\bar{P}_{spec} = \bar{P}_D \rho_o^2 \rho_s^2 \qquad 7.8$$

where:
P_D is the power applied to the surface

The mean diffuse power is equal to:

$$\bar{P}_{diff} = \bar{P}_D \rho_o^2 \rho_d^2 \qquad 7.9$$

A divergence factor can be included to account for the curvature of the earth. The resultant power is simply the addition of the powers:

$$\bar{P}_t = \bar{P}_{dir} + \bar{P}_{spec} + \bar{P}_{diff}. \qquad 7.10$$

7.9 Alternative Approach

If the beamwidth is fairly large, such as the case with a emitter detector receiver located on an aircraft, the multipath is mainly due to the energy of the reflected wave in the receiver and not the path length. The analysis follows for this type of system. As a ray of energy from a transmitter strikes a point on the earth, the reflected

Multipath

energy, reduced by the reflection coefficient, is directed towards the receiver and for specular reflection the angle of incident equals the angle of reflection. The amount of energy the receiver collects depends upon this angle. Multipath can cause errors in elevation, azimuth, range, and doppler. An analysis for one multipath ray is discussed and then the analysis is broaden to cover multiple rays. The scenario is done for a unidirectional path from a source or emitter to a receiver containing one multipath signal as shown in Figure 7-1. The first consideration is the elevation error. A mini-link budget is performed and the result is the total power received below:

$$P_T = P_A G_{Td} A_{fsd} L_{td} G_{rd} + K P_A G_{Tr} A_{fsr} L_{tr} G_{rr} \qquad 7.11$$

where:
  P_T = total received power.
  P_A = power from transmitter amplifier.
  G_{Td} = gain of transmitter antenna in the direct path.
  A_{fsd} = free space attenuation in the direct path = $10 \log c/4(PI)R_d f$.
  R_d = the direct distance between emitter and receiver.
  K = magnitude of the reflection coefficient.
  G_{rd} = gain of receiver antenna in the direct path.
  G_{Tr} = gain of transmitter antenna in the reflected path.
  A_{fsr} = free space attenuation in the reflected path = $20 \log c/4(PI)R_r f$ where: $R_r = R_d + 2h_a \sin x$ and h_a = height of the antenna.
  G_{rr} = gain of receiver antenna in the reflected path.

Since the path length difference will be small, some assumptions can simplify the equation as follows:

$$P_T = P_A A_{fs} L_t [G_{Td} G_{rd} + K G_{Tr} G_{rr}] \qquad 7.12$$

This assumes the reflected ray is directed towards the center of the antenna. If the reflected ray has a different angle due to the reflection coefficients then the equation becomes:

$$P_T = P_A A_{fs} L_t [G_{Td} G_{rd} + K G_{Tr} (G_{rr} \cos p)] \qquad 7.13$$

Note: $\qquad p = x - y$

where: $\qquad$ x = angle of the reflected ray for a smooth surface.
y = angle of the reflected ray for a specular or diffuse reflection for a rough surface.

Also, since K, G_{Tr}, and G_{rr} are dependent upon the angle of the reflected ray to the earth q, then the equation becomes:

$$P_T = P_A A_{fs} L_t [G_{Td} G_{rd} + K(q) G_{Tr}(q) (G_{rr}(q) \cos p)] \qquad 7.14$$

To determine the effect of multiple rays, the integrals with respect to the angles p and q are performed as shown:

$$P_T = P_A A_{fs} L_t [G_{Td} G_{rd} + \int \int K(q) G_{Tr}(q) (G_{rr}(q) \cos p) dq dp] \qquad 7.15$$

Multipath

If path length and the phase shift of the reflection coefficient are required in the analysis, then the following function needs to be added in:

$$e^{-j(a+b)} \qquad 7.16$$

where:

    a = phase of the reflection coefficient.
    b = phase shift of the reflected ray with respect to the direct ray

The phase shift of the reflected ray with respect to the direct ray is calculated by the following:

$$b = \frac{2\pi}{\lambda} 2 h_a \sin q \qquad 7.17$$

where:

    h_a = height of the antenna.
    q = angle of received ray with respect to ground.

For the single ray case the equation becomes:

$$P_T = P_A A_{fs} L_t [G_{Td} G_{rd} + K(q) G_{Tr}(q)(G_{rr}(q) \cos p)] e^{-j(a+b)} \qquad 7.18$$

Since the phase shift of a is generally small compared to b, and substituting for b gives:

$$P_T = P_A A_{fs} L_t [G_{Td} G_{rd} + K(q) G_{Tr}(q)(G_{rr}(q) \cos p)] e^{-j\frac{2\pi}{\lambda} 2h_a \sin q} \qquad 7.19$$

Expanding this one ray example to a multiple ray configuration is shown below:

$$P_T = P_A A_{fs} L_t [G_{Td} G_{rd} + \int\int K(q) G_{Tr}(q)(G_{rr}(q) \cos p)] \\ e^{-j(\frac{2\pi}{\lambda} 2h_a \sin q)} dp\, dq \qquad 7.20$$

This case shows a two dimensional analysis. By integrating all possibilities with respect to the angle of each multipath ray to the receiver and to the separation of the receiver to the point of reflection an estimated amplitude and angle distortion can be accomplished.

7.10 Summary

Multipath affects the desired signal by distorting both the phase and the amplitude. This can result in a lost signal or a distortion in the TOA of the desired signal. Multipath is divided into two catagories, specular and diffuse. Specular multipath generally affects the system the greatest, producing more errors. Diffuse multipath is more noise-like and is generally much lower in power. The Rayleigh criteria is used to determine if the diffuse multipath needs to be included in the analysis. The curvature of the earth can affect the analysis for very long distance multipath. One of the ways to reduce the effects of multipath is to use leading edge tracking so that most of the multipath is ignored. Some approaches in realizing the mulitpath effects include vector analysis, power summation, and an alternative approach.

7.11 References

[1] D.K. Barton, *Radar Resolution and Multipath Effects*, Radar Systems Vol. 4, Artech House, Dedham, Mass., 1975.

[2] P. Beckman, A. Spizzichino, *The Scattering of Electromagnetic Waves from Rough Surfaces*, Pergamon Press (1963).

[3] R. L. Munjal, *Comparison of Various Multipath Scattering Models*, The John Hopkins University, December 8, 1986.

[4] S.R. Bullock, "Use Geometry to Analyze Multipath Signals", *Microwaves and RF*, July 1993.

Multipath 269

Problems

1. What is the difference between glint errors and scintillation errors?

2. Which type of multipath affects the solution the most? Why?

3. What is the effect of multipath on a incoming signal if the signal is vertically polarized at the pseudo-Brewster angle?

4. What is the effect of multipath on a incoming signal if the signal is horizontally polarized at the pseudo-Brewster angle?

5. What is the criteria for determining the type of multipath to use?

6. According to the Rayleigh criteria, what is the minimum frequency that the multipath will still be considered specular for a peak height variation of 10 meters and a grazing angle of 10 degrees?

7. What is the divergence factor and how does it affect the multipath analysis?

8. How do most radars minimize multipath effects on the radar pulses?

270　　　　　　　　　　　　Transceiver System Design

9. Graphically show the resultant vector for a reference signal vector with a magnitude of 3 at an angle of 10 degrees and the smooth specular multipath signal vector with a reflection coefficient of .5 at an angle of 180 degrees?

10. What is the main difference between the vector summation approach to multipath analysis and the power summation approach? How does this approach affect the reflection coefficient?

8

Improving the System Against Jammers

Most systems are concerned about what type of jammer will prevent communications and what means are there to eliminate or reduce the effects on the reception of the desired signal. The receiver is open to reception of not only the desired signal, but all jammers that are within the receiver's bandwidth (see Figure 8-1). This can prevent the

receiver from processing the desired signal properly. Three ways are discussed in detail on how to reduce the effects of jammers. A method to protect the system against pulse or burst jammers, an adaptive filter to reduce narrow band jammers such a CW, and a jammer reduction technique called a Gram-Schmidt Orthogonalizer (GSO).

Also, in some systems, the ability for another receiver to detect the desired signal is important in the design. In order to do this properly, an understanding of the type of intercept receivers will be discussed.

Transceiver System Design

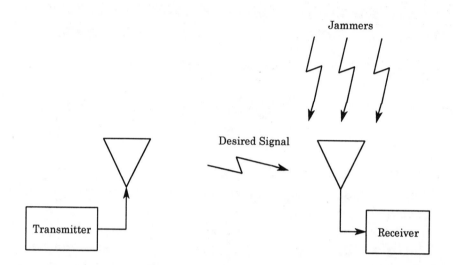

Figure 8-1 *The receiver accepts both the desired signal and jammers.*

8.1 Burst Jammer

One of the best jammers for direct sequence spread spectrum modulation is a burst jammer. A burst jammer is generally a high amplitude narrowband signal relative to the desired signal and is present for a short period of time. Typical bursts range from 0-40 dB jammer-to-signal ratio (J/S), with a duration from 0.5-1000 µs. An example of a burst jammer is shown in Figure 8-2.

The burst jammer affects the receiver as follows:

Improving the System Against Jammers

a. The high amplitude of the burst saturates the AGC (automatic gain control) amplifiers, detectors and the processor so the information is lost during the burst time and also during the recovery time of each of the devices.

b. The burst can capture the AGC. With the burst present, the AGC voltage slowly increases reducing the gain of the amplifier. When the burst is gone, the amplifier gain is small and the signal is lost until the AGC has time to respond at which time the burst comes on again.

One method of reducing the affects of a burst jammer is to use a burst clamp. A burst clamp detects the increase in power at RF and prevents the burst from entering the receiver. To determine the threshold for the power detector the previous AGC voltage is used. The process gain and the BER (bit error rate) will also be factors in determining the threshold desired. The duration thresholds will be determined by the noise spikes for a minimum and the expected changes in signal amplitude for a maximum. The burst clamp performs four functions:

Figure 8-2 An example of a burst jammer.

a. Detects the power level to determine the presence of a burst.

b. Switches out the input signal to the AGC amplifier.

c. Holds the AGC voltage to the previous value before burst occurred.

d. Tells the decoder that a burst is present.

A detector log amplifier is used to detect the power level of the burst see Figure 8-3.

The log amplifier compensates for the detector and gives a linear response of power in dB to volts out. The detected output is compared to the AGC voltage plus a threshold voltage to determine whether or not a burst is present. If a burst is present, the counter in the time-out circuit is enabled. This switches out both the IF path and the AGC path to prevent the burst from continuing through the circuit and to hold the AGC voltage at the level set before the burst occurred. The counter time-out is set for the longest expected burst and then resets the flip-flop which closes the switches to allow the AGC voltage to build up again. This prevents the burst from locking up if there is a large change in signal level or when the transmitter is turned on. The speed of the circuitry is important so that the receiver will respond to quick bursts and not allow the AGC voltage to change due to a burst. The response of the IF amplifier is slow enough that the detection circuitry has

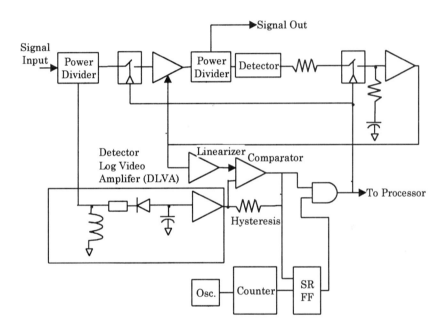

Figure 8-3 *Burst clamp used for burst jammers.*

time to respond. Some considerations when designing the circuitry are:

    a.   Linearizing and matching over temperature.

    b.   Response time of the burst clamp.

    c.   False triggering on noise.

    d.   Dynamic range of the detector.

    e.   Detector amplitude dependent on frequency.

    f.   Burst clamp saturation.

    g.   Holding AGC voltage for long burst duration.

The response time needs to be fast enough so that the burst is not in the system longer than the error correction used in system. The instantaneous dynamic range of the system (amplitude) affects the soft decision (various thresholds are used in soft decisions) which can affect the BER regardless of the process gain.

Another scheme of making a burst clamp is by using a slope detector as shown in Figure 8-4. This design uses a differentiator to detect the presence of a burst. The window comparator looks at both the rising slope and falling slope of the burst and the rest of the circuitry is basically the same as before. The advantages of this type of clamp are:

Improving the System Against Jammers

- a. The threshold is the same for all signal levels.
- b. No need for curve matching or linearizing.
- c. Insensitive to slow change in noise level and signal level.

The disadvantages are:

- a. Hard to detect the slope (detection error).
- b. Will not detect slow rise time bursts.
- c. Difficult to select the proper time constant.
- d. Slower detection process.

The placement of the burst clamp depends on the type of receiver used in the system. Most systems use a burst clamp at the IF frequency and possibly one at RF. If the system has an AGC at RF, then a burst clamp would probably be needed there also. The receiver can have either one detector at RF and use the sum of the RF and IF AGC voltages for the threshold, or use a detector at both places in the system. The latter would increase the sensitivity of the overall burst detection depending on the receiver. Most detectors have a sensitivity of around -45 dBm.

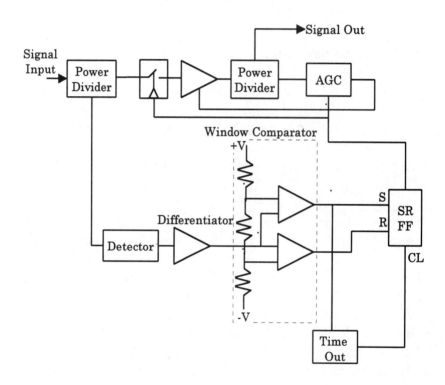

Figure 8-4 *Slope detection burst clamp.*

Improving the System Against Jammers

8.2 Adaptive Filter

An adaptive notch filter which minimizes the effect of a narrowband jammer on data links can improve the transceiver. Most adaptive tapped delay line filters function only in baseband systems with bandwidths on the order of a few MHz. In order to operate this filter over a wide bandwidth and at a high frequency, several problems must be overcome. Phase delays must be compensated for and the quadrature channels must be balanced, and therefore, the actual performance of the filter is limited by the ability to make accurate phase delay measurements at these high frequencies.

In situations where the desired signal is broadband, there is a requirement to design a filtering system that can reduce narrowband jamming signals across a very wide band at a high center frequency without an extensive amount of hardware. There are several methods that reduce the effect of the narrowband signals across a very large band.

One such method is the use of spread spectrum techniques to obtain a process gain, which improves the signal to jammer ratio. Process gain is the ratio of the spread bandwidth to the rate of information sent. However, there are limitations on the usable bandwidth, and on how slow the information can be sent in a given system. Coding techniques and interleaving can improve the system and reduce the effect of the narrowband jammer. However, the amount and complexity of the hardware required to achieve the necessary reduction limits the amount of coding that is practical in a system. But if the bandwidth is already wide, spreading the signal may be impractical and the spread spectrum techniques will be of no value.

Passive notch filters placed in-line of the receiver can be used to reduce the unwanted signals but with a large degradation in the wideband signal at the location of the notches. Also, prior knowledge of the frequency of each interferer is required and a notch for each undesired signal needs to be implemented. Since these notch filters are placed in series with the rest of the system, the overall group delay of a communication link will be altered.

Adaptive filters have been used for noise cancellation using a tapped delay line approach. The noise cancellation filter uses a separate reference input for the noise and uses the narrowband output for the desired signal. In this application, the desired signal is the wideband output, and the reference input and the signal input are the same. The reference signal goes through a decorrelation delay which decorrelates the wideband component, but the narrowband signal, because of its periodicity, remains correlated. When an adaptive filter is configured in this manner, it is known as an adaptive line enhancer (ALE).

Adaptive filters, configured as ALEs, have been used to reduce unwanted signals; however, they have been limited to relatively narrowband systems and at low frequencies. Adaptive filters can be used in broadband systems with a high carrier frequency provided certain modifications are made to the system.

Improving the System Against Jammers 281

8.3 Digital Filter Intuitive Analysis

A finite impulse response (FIR) digital filter is shown in Figure 8-5. The signal enters the tapped delay line and each output of the taps is multiplied by the weight value,

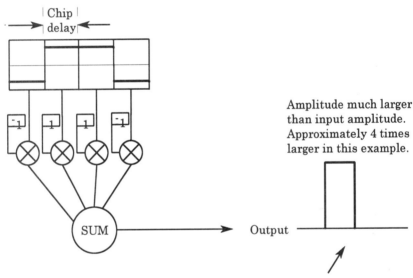

Figure 8-5 FIR matched filter.

a number, and then they are all summed together. One way to look at this is that each tap is moving up and down according to the input signal and time. Therefore, the sum is equal to a point on the sine wave in time. This could be phased shifted, but that is not a problem since analog filters have phase shifts (delays). At another point in time, the signal levels are different in the taps, however, the multiplication and resulting sum equals another point on the sine wave. This continues to happen until the output is the sinewave or the input sinewave with a delay. Note that other frequencies will not add up correctly with the given coefficients and will be attenuated.

An adaptive filter uses feedback to adjust the weights to the correct value for the input sine wave compared to the random signal, and in an ALE, the sine wave is subtracted from the input signal to result in just the wideband signal output. Note that there is a decorrelation delay in the ALE which delays the wideband signal such that the autocorrelation of the wideband signal is small with a delayed version of itself and the narrowband signal has high autocorrelation with a delayed version of itself. This technique is used to reduce narrow band jamming in a wide band spread spectrum system.

8.4 Basic Adaptive Filter

A block diagram of a basic adaptive filter configured as an adaptive line enhancer (ALE) is shown in Figure 8-6. The wideband input d_k consists of a narrowband signal s_k plus a wideband signal or noise n_k. The composite signal is split

Improving the System Against Jammers

and one channel is fed to a decorrelation delay indicated by Z^{-D} which decorrelates the wideband signal or noise. The other goes to a summing junction. The output of the decorrelation delay is delivered to a chain of delays and the output of each delay $X_k(i)$ is multiplied by its respective weight values $W_k(n)$. The weights are drawn with arrows to indicate that they are varied in accordance to the error feedback e_k. The outputs of all the weights are summed together producing the estimated narrowband spectral line y_k. This narrowband signal is subtracted from the wideband input d_k to produce the desired wideband signal output, which is also the error e_k. The name, adaptive line enhancer, indicates that the desired output is the narrowband signal y_k. However, for use in narrowband signal suppression, the wideband or error signal e_k is the desired output.

Adaptive line enhancers are different from fixed digital filters because they can adjust their own impulse response. They have the ability to change their own weight coefficients automatically using error feedback, with no *a priori* knowledge of the signal or noise. Because of this ability, the adaptive line enhancer is a prime choice in jammer suppression applications where the exact frequency is unknown. The ALE generates and subtracts out the narrowband interferer leaving little distortion to the wideband signal. Also, one adaptive line enhancer can take out more than one narrowband interference signal at a time, in a given bandwidth. The adaptive filter scheme has the ability to adapt in frequency and amplitude to the interferer in a specified bandwidth so exact knowledge of the interferer is not necessary. If the interferer changes in

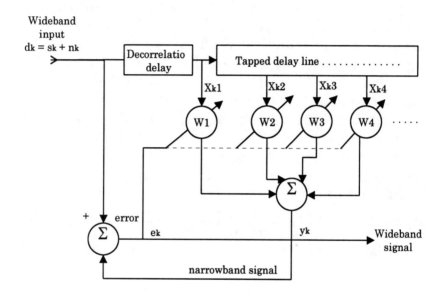

Figure 8-6 *Basic adaptive filter configured as an ALE.*

frequency in the given band, the filter will automatically track the change and reduce the narrowband signal.

8.5 LMS Algorithm

The adaptive filter works on the principle of minimizing the mean-squared-error $E[e_k^2]$ using the least-mean-squared (LMS) algorithm [4] see Figure 8-7.

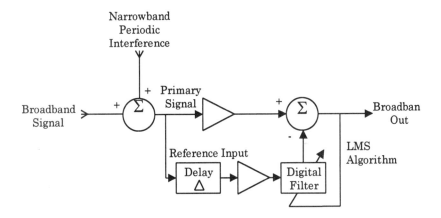

Figure 8-7 Adaptive filter using LMS algorithm.

The input to the filter is:

$$d_k = s_k + n_k \quad \text{(scalar)} \qquad 8.1$$

s_k = narrowband signal (scalar)

n_k = noise or wideband signal (scalar)

Basically the ALE is designed to minimize the mean-squared error (MSE).

$$\begin{aligned} MSE &= E[e_k^2] = E[(d_k - W_k^T X_k)^2] \\ &= E[d_k^2 - 2d_k W_k^T X_k + (W_k^T X_k)^2] \\ &= E[d_k^2] - 2E[d_k X_k^T]W_k + E[W_k^T X_k X_k^T W_k] \quad 8.2 \end{aligned}$$

e_k = error signal (scalar)

W_k^T = weight values transposed (vector)

X_k = the tap values (vector)

By substituting the autocorrelation and crosscorrelation function definitions gives:

$$MSE = E[d_k^2] - 2r_{xd}W_k^T + W_k R_{xx}^T W_k \qquad 8.3$$

Improving the System Against Jammers

$$R_{xx} = E[X_k X_k^T] = \text{autocorrelation matrix}$$

$$r_{xd} = E[d_k X_k^T] = \text{crosscorrelation matrix}$$

$$X_k^T W_k = X_k W_k^T \qquad 8.4$$

To minimize the MSE, the gradient, ∇_w, is found and set equal to 0:

$$\nabla_w E[e_k^2] = 2 R_{xx} W_k - 2 r_{xd} = 0 \qquad 8.5$$

Solving for W which is the optimal weight value gives:

$$W_{opt.} = R_{xx}^{-1} r_{xd} \qquad 8.6$$

The weight equation for the next weight value is:

$$W_{k+1} = W_k - u \nabla_w E[e_k^2] \qquad 8.7$$

In the LMS algorithm, e_k^2 itself is the estimate of the mean squared error. Because the input signal is assumed to be ergodic, the expected value of the squared error can be estimated by a time average. The weight equation then becomes:

$$W_{k+1} = W_k - u \nabla_w(e_k^2)$$
$$= W_k + 2 u e_k X_k \qquad 8.8$$

This is known as the LMS algorithm and is the method that is used to update the weights. The new weight value W_{k+1} is produced by summing the previous weight value to the product of the error e_k, times the tap value X_k, times a scale factor u. The u value determines the convergence rate and stability of the filter. It has been shown that under the assumption of ergodicity, the weights converge to the optimal solution $W_{opt.}$.

8.6 Digital/Analog ALE

In digital adaptive filters, it is assumed that the time for the error signal to be generated and processed to update the weight values is less than a clock cycle delay of the digital filter clock. However, since part of the feedback loop is analog, delay compensation is required. A quadrature method is used in the frequency conversion processes to allow ease of tuning across a very wide band of operation. This also provides a wider instantaneous bandwidth.

Unless the signal can be digitized at the RF frequency, a digital/analog combination needs to be implemented. When adaptive line enhancers are used at high frequencies, there must be a downconversion of the RF signal before it can be processed and digitized by the digital filter. This is done using a local oscillator (LO) and a mixer to downconvert the signal to baseband for processing by the digital filter shown in Figure 8-8.

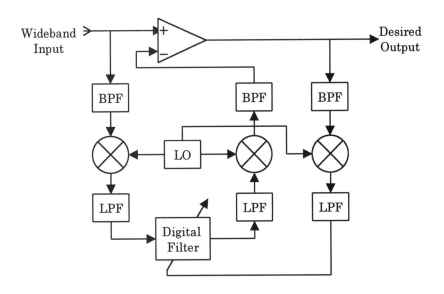

Figure 8-8 *Analog/Digital adaptive filter.*

A bandpass filter provides a coarse selection of the band for processing. The actual processing bandwidth of the signal is limited to the clock frequency of the digital filter due to the Nyquist criteria to prevent aliasing. After the tone is generated by the digital filter, it is upconverted, using a mixer and the same LO, to the RF frequency. The signal is fed through a bandpass filter to eliminate the unwanted sideband produced in the upconversion process and is then subtracted from the composite signal. The final output is split and used for the error feedback signal, which is downconverted in the same manner as the reference signal, see Figure 8-9.

Large bandwidths can be achieved by using a synthesizer to vary the LO frequency and select a band where narrowband interference is a problem. This would not produce a large instantaneous bandwidth, but it would allow control over a large band of frequencies with relatively little hardware. However, sidebands need to be filtered to process the signal correctly. One way to accomplished this is to use tunable filters as shown in Figure 8-9.

A quadrature up/down conversion scheme can be used to retrieve the desired signal as shown in Figure 8-10. This configuration allows the LO to be positioned in the center of the band of interest and eliminates the need for tunable filters. The filter is similar to the system shown in Figure 8-8, except that in this configuration two digital filters and two channels in quadrature are required. A wide bandpass filter can be used in the IF section to select the entire band that the adaptive filter will tune across. The LO or synthesizer selects the band of interest as shown in Figure 8-10. The I and Q signals are lowpassed filtered. Both

Improving the System Against Jammers 291

sidebands are downconverted to baseband with an in-phase signal from the oscillator for the 'I' channel and a quadrature-phase signal from the oscillator for the 'Q' channel. This method allows twice the processing bandwidth with a given clock frequency because it utilizes both sidebands in the process. There is a digital filter for both the I and Q channels which generate the respective tones. During the upconversion of the digital filter outputs, each channel produces both bands on each side of the carrier. However, the phase relationship of the bands provides cancellation of the unwanted sidebands when the signals are summed together, so that only the desired sidebands are generated as shown in Figure 8-10.

This behaves the same as a quadrature transmitter/receiver combination for sending two different signals and receiving them.

Two signals are quadrature downconverted with an in-phase (I channel) and a 90 degree phase shift (Q channel) to give:

I-channel = $\cos(\omega_1 t) + \cos(\omega_2 t)$ 8.9

Q-channel = $\cos(\omega_1 t - 90) + \cos(\omega_2 t - 90)$ 8.10

where:

ω_1 = a frequency located in band 1

ω_2 = a frequency located in band 2

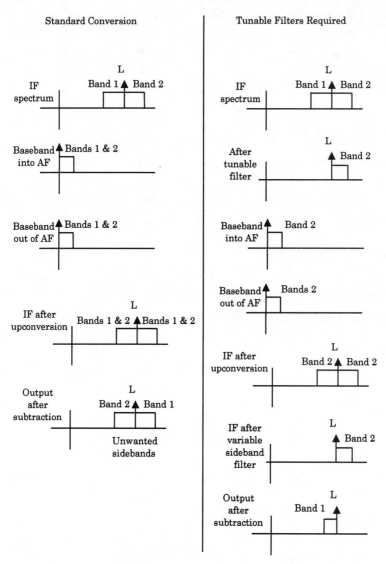

Figure 8-9 *Tunable filters required to eliminate unwanted sidebands.*

Improving the System Against Jammers

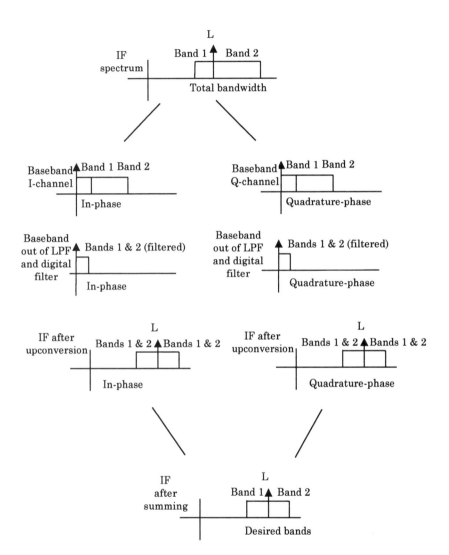

Figure 8-10 *Quadrature methods eliminate unwanted sidebands.*

The two digital filters generate these signals and deliver them to the upconversion process. The same LO is used for both the up and down conversion. The in-phase LO is mixed with the I channel and the 90 degree phase shifted LO is mixed with the Q channel. The results are as follows, which include the sum and difference terms:

$$\text{I-channel} = \cos(\omega_0 + \omega_1 t) + \cos(\omega_0 - \omega_1 t)$$
$$+ \cos(\omega_0 + \omega_2 t) + \cos(\omega_0 - \omega_2 t) \qquad 8.11$$

$$\text{Q-channel} = \cos(\omega_0 t - 90 + \omega_1 t - 90)$$
$$+ \cos(\omega_0 t - 90 - \omega_1 t + 90)$$
$$+ \cos(\omega_0 t - 90 + \omega_2 t - 90)$$
$$+ \cos(\omega_0 t - 90 - \omega_2 t + 90)$$
$$= \cos(\omega_0 + \omega_1 t - 180) + \cos(\omega_0 - \omega_1 t)$$
$$+ \cos(\omega_0 + \omega_2 t - 180) + \cos(\omega_0 - \omega_2 t) \qquad 8.12$$

Amplitudes have been neglected to show phase response only. When summed together the net result is:

$$2\cos(\omega_0 - \omega_1 t) + 2\cos(\omega_0 - \omega_2 t) \qquad 8.13$$

Improving the System Against Jammers

The unwanted sidebands are eliminated due to the net 180 degree phase shift as shown in Figure 8-10. If the desired frequency occurred on the high side of the LO, there would have been a sign change in the solution above. Since the unwanted sidebands are eliminated due to the quadrature scheme, tunable filters are no longer required for the upconversion process.

Another problem to overcome in this implementation is the LO bleedthrough when using a non-ideal mixer. LO bleedthrough is the amount of local oscillator signal appearing on the output of a mixer due to imbalance of a double balanced mixer. Since the idea is to eliminate jamming signals in the passband, the LO bleedthrough will appear as another jamming signal which is summed with the original signal. Designing a well-balanced mixer will help the problem, but there is a limit on how well this can be done. One way to reduce LO bleedthrough is to couple the LO and phase shift the coupled signal so it is 180 degrees out of phase with the LO bleedthrough and exactly the same amplitude, and then sum the coupled signal with the composite signal thus reducing the unwanted LO signal. However, since the synthesizer is changing in frequency and the group delay for the LO bleedthrough is not constant, this becomes very difficult to accomplish. An alternative is to use an intermediate frequency (IF) with a fixed LO to upconvert from baseband to the IF and then use the synthesizer to mix up to the desired frequency. By doing this, the mixers at the lower frequency band provide better LO isolation so the fixed LO bleedthrough is greatly reduced and the synthesizer bleedthrough is out of band and can be easily filtered.

8.7 Wideband ALE Jammer Suppressor Filter

The wideband ALE jammer suppressor filter is shown in Figure 8-11. The adaptive filter is connected in parallel with the communication system and the only components in line with the system are three couplers and one amplifier. The group delay through the amplifier is constant across the band of interest with a deviation of less than 100 ps. The amplifier is placed in the system to isolate the reference channel from the narrowband channel. This prevents the narrowband signal from feeding back into the reference channel.

The wideband high-frequency composite signal is split using a −10 dB coupler, amplified, and downconverted to an IF using an adjustable synthesizer. This provides the reference signal for the digital filter. The signal is filtered, amplified, and quadrature downconverted to baseband where the quadrature signals are filtered and amplified and provide the reference signals to the digital filters. Elliptic lowpass filters are used to achieve fast rolloffs and relatively flat group delay in the passband. The digital filters produce the estimated narrowband signals. The outputs of the digital filters are lowpass filtered, amplified and are quadrature upconverted to the IF band and summed together to eliminate the unwanted sidebands. The signal is filtered and upconverted by the synthesizer to the desired frequency and then filtered, amplified and subtracted from the composite signal to eliminate the narrowband signal. The output provides both the desired wideband signal and the error. This signal is split and the

Improving the System Against Jammers

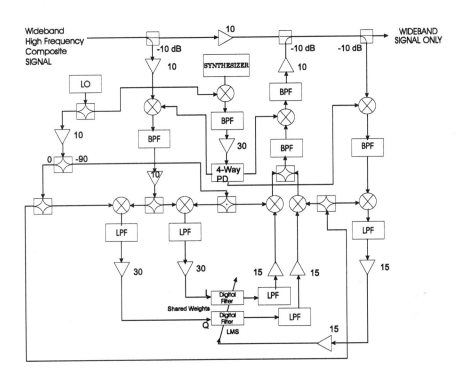

Figure 8-11 *Wideband ALE jammer suppressor filter.*

in-phase error is downconverted in two stages to baseband to update the filter weights as shown in Figure 8-11.

8.8 Digital Circuitry

The digital filter digitizes the reference input using A/D converters and feeds this signal through a decorrelation delay. This decorrelates the wideband signal and allows the narrowband signal, because of its periodicity, to remain correlated. The delayed signal is fed to a tapped delay line containing 16 different time-delayed signals or taps that are multiplied by the error feedback and then accumulated to update the weight values. The analog error signal is single-bit quantized and scaled by adjusting the u value. The u value determines the filters sensitivity to the error feedback. The 16 taps are then multiplied by these new weight values and converted to analog signals where they are summed together to form the predicted narrowband output. A selectable delay was incorporated in the design to provide adjustment for the time delay compensation. This was necessary due to the delay through the digital and analog portions of the filter. When the filter generates the predicted narrowband signal for cancellation, the error produced needs to update the portion of signal that caused the error. If there is delay in this path, then the error will be updating a different portion of the signal than the part that generated the error. Since the analog portion produces delay that is not quantized with regards to a certain number of clock cycles, a variable delay was designed to select small increments of a clock cycle. This allows for better resolution in selecting the correct

Improving the System Against Jammers

compensation delay. A rotary switch, mounted on the board, is provided for adjusting the delay for each digital filter.

8.9 Simulation

A limited simulation of the ALE system was performed due to the complexity and large number of variables contained in ALE system definition. Since simulation time is dependent on the sample rate of the computer clock, a baseband representation was used. The amount of computer time to do the simulation at high frequencies is impractical. Existing models were used to form the desired circuit. The frequency is swept across a 30 MHz band and the output power is measured and displayed. The results are shown in Figure 8-12. This shows the cancellation across the instantaneous bandwidth. The filter is shown to be unstable at the filter edges. This is a result of phase distortion and aliasing effects. The spikes at DC are a result of only using a 16 tap filter. The low frequency components cannot be resolved in the tap delay line. The cancellation achieved in the simulation is greater than 40 dB except around DC.

8.10 Results

The adaptive filter cancellation bandwidth is shown in Figure 8-13 which is similar to the results obtained in the simulation. This shows the response of the filter as a CW tone is swept across the selected band. The bandwidth is less than the simulation in order to suppress the phase distortion at the band edge. The center of the band is the response of the filter at DC. Since the ALE is AC-coupled in the hardware, it cannot respond to DC or low frequencies, which results in no cancellation of the signal. This differs from the simulation because the simulation is DC-coupled. The amount of cancellation across the band is approximately 10 dB, with up to 30 dB at certain frequencies. This can be compared to the 40 dB cancellation across the band achieved in the simulation. Quadrature imbalance is a major factor in the amount of suppression of the narrowband tone and is difficult to measure at high frequencies using frequency conversion processes. The quadrature balance was estimated, and the hardware tuned to achieve maximum performance. The synthesizer can be tuned to select a specified frequency to achieve maximum cancellation of that frequency across a given bandwidth. The amount of cancellation is dependent on many factors such as I-Q balance for both phase and amplitude, phase linearity across the band, and stability and noise in the system. The cancellation for a single interferer is shown in Figure 8-14. The response shows a single interferer with and without the ALE in the system. The single tone is suppressed by approximately 20 dB. ALE's can suppress more than one interferer at a time as shown in Figure 8-15. Generally, there is a degradation in performance in the amount of suppression with respect to the number of interferers. Also, more tones produce more spurious responses in the mixing processes in the system.

Improving the System Against Jammers 301

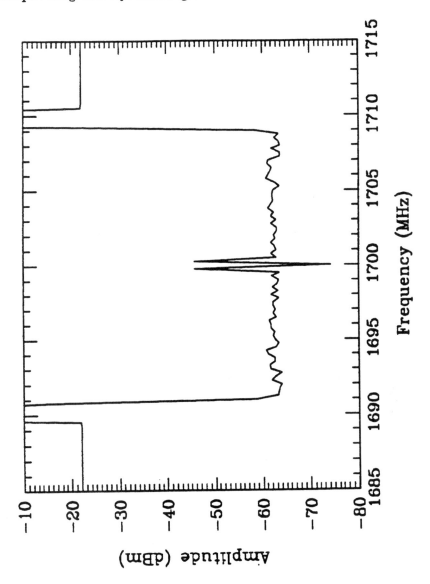

Figure 8-12 *Simulation results using Systid*

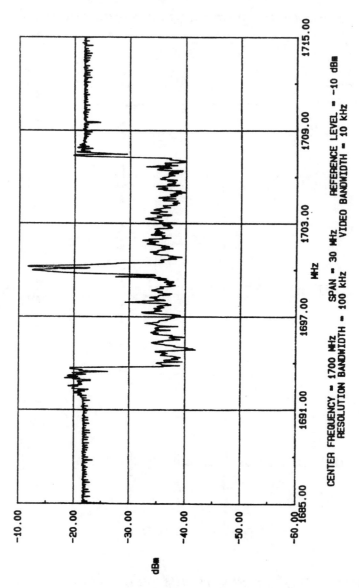

Figure 8-13 *Adaptive filter cancellation bandwidth.*

Improving the System Against Jammers

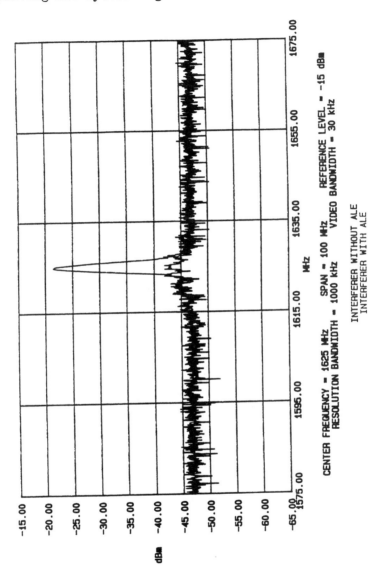

Figure 8-14 *Single interferer suppression results.*

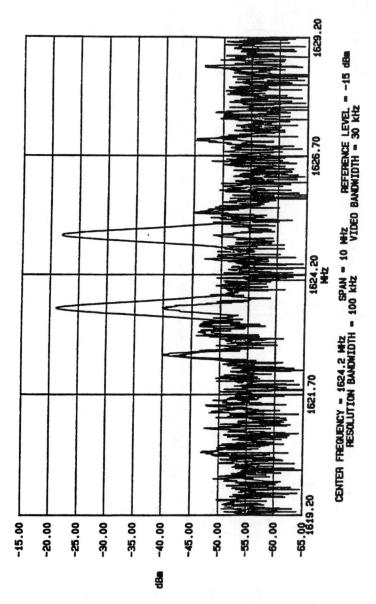

Figure 8-15 Double interferers suppression results

Improving the System Against Jammers

8.11 Gram-Schmidt Orthogonalizer (GSO)

Signals can be represented by the sum of weighted orthonormal functions as shown below:

$$S_1(t) = aX_1(t)$$
$$S_2(t) = bX_1(t) + cX_2(t)$$

8.14

$$\cdot$$
$$\cdot$$
$$\cdot$$

where $X_1(t)$, $X_2(t)$ ... are orthonormal functions and a,b,c... are the weighting coefficients.

The constant 'a' is simply the magnitude of $S_1(t)$ since the magnitude of $X_1(t)$ by definition is unity. To solve for 'b', the second equation is multiplied by $X_1(t)$ and integrated:

$$\int S_2(t)X_1(t)dt = \int bX_1(t)X_1(t)dt + \int cX_2(t)X_1(t)dt$$

8.15

On the right side of the above equation, the second term is equal to zero (the inner product of two orthonormal functions = 0) and the first term is equal to 'b' (the inner product of a orthornormal function with itself = 1). Solving for 'b' produces:

$$b = \int S_2(t)X_1(t)dt$$

8.16

which is the projection of $S_2(t)$ on $X_1(t)$.

The constant 'c' is determined by the same procedure except that $X_2(t)$ is used in place of $X_1(t)$ when multiplying the second equation. Therefore:

$$c = \int S_2(t) X_2(t) dt \qquad 8.17$$

which is the projection of $S_2(t)$ on $X_2(t)$. Note that $X_2(t)$ is not generally in the direction of $S_2(t)$.

The first orthonormal basis function is therefore defined as:

$$X_1(t) = S_1(t)/a \qquad 8.18$$

where a is the magnitude of $S_1(t)$.

The second orthonormal basis function is derived by subtracting the projection of $S_2(t)$ on $X_1(t)$ from the signal $S_2(t)$ and dividing by the total magnitude:

$$X_2(t) = [S_2(t) - bX_1(t)]/M \qquad 8.19$$

where M is the magnitude of the resultant vector:

$$|S_2(t) - bX_1(t)| \qquad 8.20$$

A phasor diagram is provided in Figure 8-17 to show the projections of the vectors.

Improving the System Against Jammers

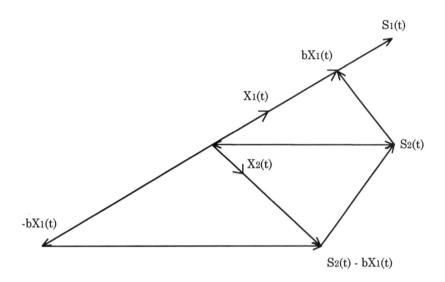

$X_1(t) = S_1(t)/a$ = first orthonormal function.

$X_2(t) = [S_2(t) - bX_1(t)]/M$ = second orthonormal function.

where M is the magnitude of the resultant vector.

Figure 8-17 *Orthonormal vectors.*

8.12 Basic GSO

A basic GSO system is shown in Figure 8-18. The weight (w_1) is chosen so that the two outputs V_o and W_o are orthogonal, that is the inner product $<V_o, W_o> = 0$. This gives the result:

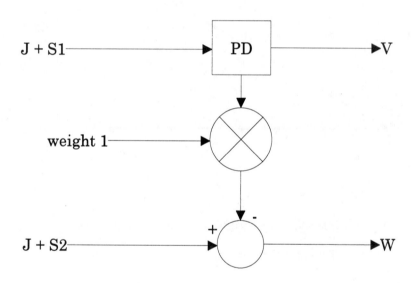

Figure 8-18 *Basic GSO.*

Improving the System Against Jammers

$$\langle J+S_1, J+S_2-w_1(J+S_1)\rangle = 0 \qquad 8.20$$

$$\langle J,J\rangle(1-w_1)-w_1\langle S_1,S_1\rangle = 0 \qquad 8.21$$

$$w_1 = \frac{|J|^2}{|J|^2+|S_1|^2} = \frac{1}{1+p} \qquad 8.22$$

$$p = \frac{1-w_1}{w_1} = \frac{|S_1|^2}{|J|^2} \qquad 8.23$$

Note the following assumptions:

a. Same J in both inputs.
b. J, S_1, S_2 are orthogonal.
c. $|J|^2$, $|S_1|^2$, $|S_2|^2$ are known.
d. p<<1. This means the jammer is much larger than the signal S_1.

The outputs are:

$$V_o = J + S_1 \qquad 8.24$$

$$\begin{aligned}W_o &= J + S_2 - w_1(J+S_1) \\ &= J(1-w_1) + S_2 - w_1 S_1 \\ &= J(p/(1+p)) + S_2 - S_1(1/(1+p))\end{aligned} \qquad 8.25$$

Since $p \ll 1$ then:

$$W_o = J_p + S_2 - S_1 = |S_1 - S_2|^2 \qquad 8.26$$

which shows that the jammer signal has been attenuated in the W_o output.

Suppose only a jammer exists in one of the inputs S_1 and jammer plus signal is in S_2 as shown in Figure 8-19. Taking the projection of S_2 on the orthonormal function ($Q_1 = S_1/|S_1|$) provides the amount of jammer present in S_2:

$$b = \langle S_2, \frac{S_1}{|S_1|} \rangle \qquad 8.27$$

This scalar quantity multiplied by Q_1 produces the jammer vector bQ_1. Subtracting the jammer vector from S_2 gives the amount of signal present (S_p) in S_2:

Improving the System Against Jammers

$$S_p = S_2 - bQ_1 = S_2 - \langle S_2, \frac{S_1}{|S_1|}\rangle \frac{S_1}{|S_1|} \qquad 8.28$$

Thus, the jammer is eliminated. In the implementation of these systems, the $|S_1|$ values are combined in a scale factor k where:

$$k = 1/|S_1|^2$$

The final result is then:

$$S_p = S_2 - k\langle S_2, S_1\rangle S_1 \qquad 8.29$$

The constant k is incorporated in the specification of the weight value or in the integration process during the generation of the weights in an adaptive system.

8.13 Adaptive GSO Implementation

If a system uses an omni antenna and assumes jammer only for S_1 (since the jammer is much larger in amplitude than the desired signal), and a directional antenna for signal plus jammer for S_2 (since the antenna will be pointed towards the desired signal), the example above will apply.

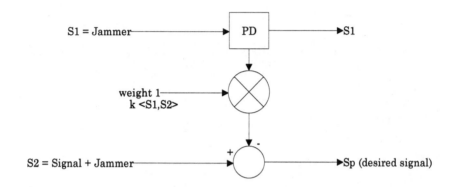

Figure 8-19 *Simple jammer suppressor.*

An adaptive filter configured as an adaptive noise canceler can be used as a GSO jammer suppressor (see Figure 8-20). This shows a quadrature system with separate I and Q outputs and separate weight generators. The error signal is produced by subtracting the weighted reference input signal (the received signal from the omni antenna) from the signal received from the directional antenna:

$$e = S_{dir} - w_1 S_{omni} \qquad 8.30$$

Improving the System Against Jammers 313

The squared error is therefore:

$$e^2 = S_{dir}^2 - 2w_1 S_{dir} S_{omni} + w_1^2 S_{omni}^2 \qquad 8.31$$

The mean squared error is:

$$MSE = \overline{e^2} = w_1^2 \overline{S_{omni}^2} - 2w_1 \overline{S_{dir} S_{omni}} + \overline{S_{dir}^2} \qquad 8.32$$

Taking the gradient of the MSE and setting this equal to 0, the optimum weight value can be solved:

$$2w_1 \overline{S_{omni}^2} - 2\overline{S_{dir} S_{omni}} = 0 \qquad 8.33$$

$$w_1(opt) = \overline{S_{dir} S_{omni}} / \overline{S_{omni}^2} \qquad 8.34$$

Since the error output is the desired signal output, then:

$$e = S_{dir} - (\overline{S_{dir} S_{omni}} / \overline{S_{omni}^2}) S_{omni} \qquad 8.35$$

The inner product is defined as:

$$\langle X_1, X_1 \rangle = \int X_1^2(t) dt = E[X_1^2(t)] = \overline{X_1^2(t)} = |X_1(t)|^2 \qquad 8.36$$

$$\langle X_1, X_2 \rangle = \int X_1(t) X_2(t) dt = E[X_1(t) X_2(t)] = \overline{X_1(t) X_2(t)} \qquad 8.37$$

Therefore the error can be expressed as:

$$e = S_{dir} - \left(\frac{\langle S_{dir}, S_{omni}\rangle}{|S_{omni}|^2}\right) S_{omni} \qquad 8.38$$

The second term on the right side of the above equation is the projection of S_{dir} on the orthonormal function $(S_{omni}/|S_{omni}|)$ times the orthonomal function. This determines the amount of jammer present in S_{dir}. This result is then subtracted from S_{dir} to achieve the desired signal, e, and eliminate the jammer.

The LMS algorithm assumes that the gradient of the MSE can be estimated by the gradient of the squared error, which turns out to be twice the error times the reference.

8.14 Intercept Receiver Comparison

Some receivers are designed to intercept the transmissions of an unknown transmitter. These types of receivers are called intercept receivers. There are many types of intercept receivers that are used to perform this task. The following lists some types of intercept receivers, often referred to as electronic countermeasures (ECM) receivers

Improving the System Against Jammers

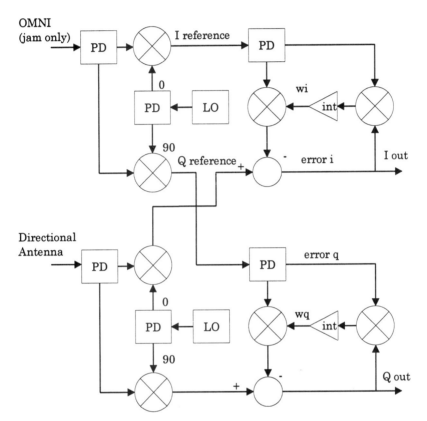

Figure 8-20 *Quadrature adaptive system.*

that are designed to listen to broadcasts from other sources (see Figure 8-21). Included are the disadvantages and advantages to using them. This is not intended to be a comprehensive list, but is provided to give a general idea of what type of receivers could detect the desired signal.

a. Crystal Video Receiver: This type of receiver uses a crystal video detector circuit that detects to see if there is a signal present. This is generally for narrowband detection (to limit the noise into the detector) unless tunable filters are employed.

   1. The advantages of a crystal video receiver are small in size, simple to design, and is less expensive compared to other types of intercept receivers.

   2. The disadvantage of a crystal video receiver is that it does not have a broad frequency detection ability unless tunable filters are used in the front end which are complex and expensive.

b. Instantaneous Frequency Measurement (IFM) (delay line discriminator receiver): This type of receiver usually consists of a delay line discriminator design to detect one single frequency.

   1. The advantages of a IFM receiver are high frequency resolution detection capability and the ability to detect wide frequency bandwidths.

Improving the System Against Jammers 317

2. The disadvantage of a IMF receiver is that the receiver cannot handle multiple frequencies at the same time.

c. Scanning Superheterodyne Receiver: This receiver uses a standard superheterodyne receiver with a tuning oscillator to cover the wide frequency bandwidth. However, this is not instantaneous and is dependent on the sweep time of the oscillator. This is sometimes referred to as a spectrum analyzer receiver.

1. The advantages of a scanning superheterodyne receiver are good frequency resolution detection capability and coverage of a wide frequency dynamic range.

2. The disadvantages of a scanning superheterodyne receiver are that it does not cover the band instantaneously, it is dependent on scan time, and the spurs that are generated in the scanning process result in false detections of the signal.

d. Instantaneous Fourier Transform (IFT) (Bragg Cell, Acousto-Optic Device): This type of receiver uses an acousto-optic device such as a Bragg Cell. Simply stated, this receiver uses optic rays which project on a surface at different points dependent on the incoming frequency. The frequency can then be determined on the location of the ray.

1. The advantages of an IFT are good frequency resolution detection capability, multiple frequency handling, and instantaneous wideband coverage.

2. The disadvantages of an IFT are the small dynamic range (20-30 dB) and the size of the receiver.

e. Microscan Receiver (extremely fast scanner): This type of receiver, often referred to as a chirped receiver, uses surface acoustic waves (SAW) device that is excited by an impulse input signal. Since all frequencies are contained in an impulse, these frequencies are mixed with the incoming signal and delayed by means of acoustic waves and the output is time dependent on the resultant frequencies. This is basically an extremely fast scanning receiver.

1. The advantages of a microscan receiver are wideband instantaneous frequency coverage, good frequency sensitivity, and very small size.

Improving the System Against Jammers

2. The disadvantages of a microscan receiver are limited dynamic range capability (30-40 dB) and the pulse information is lost. The dynamic range reduction is due to the sidelobes that are generated in the process.

f. Channelized Receiver: This type of receiver uses multiple frequency channels that provide true instantaneous frequency processing. This is basically several receivers fused together with each receiver containing bandwidths to cover the total desired broad bandwidth.

1. The advantages of a channelized receiver are good frequency resolution detection capability, detectability of multiple frequencies, and a wide frequency instantaneous dynamic range.

2. The disadvantages of a channelized receiver are the size and cost.

8.15 Summary

Adaptive filters can be configured as adaptive line enhancers to suppress undesired narrowband signals. When these filters are used at high frequencies and over large bandwidths, modifications need to be made. The

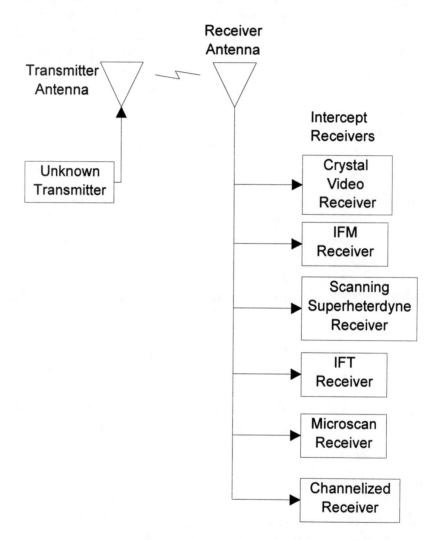

Figure 8-21 Intercept receivers for signal detection.

Improving the System Against Jammers 321

time delay through portions of the system needs to be compensated and is accomplished in the digital filter by using a tap value to generate the signal and a delayed tap value to update the weights. Also, for small variations in delay, the clock is modified in the digital filter. Once the delay is set, constant delay over the band of operation is important to maintain for proper operation of the filter. A variable synthesizer is used in the design to achieve a wide operational bandwidth for the ALE. This allows the filter to be positioned across a given bandwidth with minimal hardware. A quadrature scheme is used to eliminate filtering constraints and provide twice the processing bandwidth for the ALE. However, the performance of the system relies on the quadrature channels being balanced in both phase and amplitude. The LO bleedthrough problem is reduced by using a double downconversion scheme since the isolation of the LO signal is greater for lower frequencies. The isolation for the conversion process is established in the first low frequency mixer because the LO signal for the second mixer lies outside of the passband and can be filtered.

Gram-Schmidt Orthogonalizers can be used to reduce the effects of jamming signals. One of the assumptions in this approach is that the jammer signal level is much higher than the desired signal level. The basic GSO uses two inputs with one of the inputs containing more signal than jammer. This applies to having two antennas with one of the antennas directed towards the signal providing higher signal power. The error signal for feedback in updating the weight value is produced by subtracting the weighted reference input signal from the received signal which contains the higher level of desired signal. When the weight has converged, then the jamming signal is suppressed.

Spread spectrum systems can be used to reduce the detectability of the desired signal. These type of systems are called low probability of intercept (LPI) systems. A lot of research is performed to design a better intercept receiver, and then to design a better LPI receiver, and the race goes on.

8.16 References

[1] Bernard Widrow, Samuel D. Stearns, *Adaptive Signal Processing*, N.J.: Prentice-Hall, pp. 354-361, 1985.

[2] Bernard Widrow, John R. Glover, JR., John M. McCool, John Kaunitz, Charles S. Williams, Robert H. Hearn, James R. Zeidler, Eugene Dong, JR., and Robert C. Goodlin, "Adaptive Noise Cancelling: Principles and Applications," *Proc. IEEE*, Vol. 63, p. 1693, Dec. 1975.

[3] F.A. Bishop, R.W. Harris and M.C. Austin, "Interference Rejection Using Adaptive Filters," *Electronics for National Security Conference Proceedings*, pp.10-15, September 1983.

[4] Simon Haykin, *Communication Systems*, New York: Wiley & Sons, pp. 250-251, 1978.

[5] Scott R. Bullock, "High Frequency Adaptive Filter", *Microwave Journal*, Sept. 1990.

Problems

1. What is meant by capturing the AGC of a system?

2. What would be a good pulse frequency for a burst jammer given an AGC response time of 1µs?

3. What is the main difference between a digital FIR filter and an adaptive filter?

4. Why is either filtering or quadrature method required to operate the adaptive filter in the RF world?

5. What is the u value in the LMS algorithm represent?

6. What is the result of increasing the u value on convergence time, stability and steady state accuracy?

7. What is the biggest difference in a digital system compared to an analog system?

8. Why is the assumption that the jammer is the only signal present in the omni-directional antenna of a GSO jammer suppression filter a good assumption?

9. When might the assumption that the jammer is the only signal present in the omni-directional antenna

Improving the System Against Jammers 325

of a GSO jammer suppression filter be a bad assumption?

10. What is the best detection receiver if cost, size, and complexity is not a criteria and why?

9

Global Navigation Satellite Systems

The last few years there has been real interest in the commercialization of Global Navigation Satellite Systems (GNSS) which is often referred to as Global Positioning Systems (GPS) in various applications. A GPS system uses spread spectrum signals, BPSK, emitted from satellites in space for position and time determination. Until recently, the use of GPS was essentially reserved for military use. Now there is great interest in using GPS systems for navigation of commercial aircraft. The FAA is looking into the possibility of using a wide area augmentation system (WAAS) to cover the whole United States with one system. There are also applications in the automotive industry, surveying, and personal use for hikers and recreationalists. Due to the increase in popularity of GPS, and since it is a spread spectrum communication system using BPSK, a brief introduction is included in this text.

9.1 Satellite Transmissions

The NAVSTAR GPS satellite transmits a direct sequence BPSK signal at a rate of 1.023 Mbps using a code length of 1023 bits. The time between code repetition is 1 ms. This is known as the coarse/acquisition (C/A) code which is used by the military for acquisition of a much longer precision code (P-code). The commercial industry uses the C/A code for the majority of the applications. There are 36 different C/A codes that are used with GPS which are generated by modulo-2 adding the C/A code with a different delayed version of the same C/A code, see Figure 9-1. Therefore, there are 36 different time delays to generate the 36 different C/A codes. There are only 32 codes that are used presently for satellite operation and each is assigned a PRN number.

There are two frequencies that are transmitted by the satellite. The C/A code uses only one of these frequencies, L1, for transmission. The frequency of L1 is 1.57542 GHz. Therefore, there are 1540 carrier cycles of L1 for each C/A chip. The frequency of L2 is 1.2276 GHz and does not contain the C/A code. The P-code, which is used by the military, is transmitted on both frequencies. The P-code is transmitted in quadrature (90 degrees) with the C/A code on L1. The carrier phase noise for 10 Hz one-sided noise BW is 0.1 radian RMS for both carriers and the in-band spurious signals are less than 40 dBc.

The antenna used in the satellite for transmissions has a 3 dBi gain and is linear, right-hand circularly polarized.

Global Positioning System   329

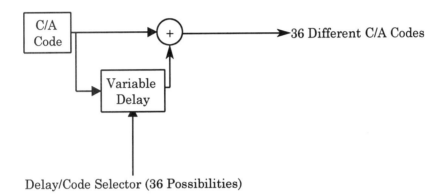

Figure 9-1 *Generating 36 different C/A codes.*

The group delay deviation for the transmitter is within ± 3 ns, 2 sigma.

9.2 Data Signal Structure

The satellite navigation data is sent out at a 50 bits per second rate. As a general rule, the information bandwidth should be at least 1 kHz. The data is sent out in 25 full frames with 5 subframes in each frame. Each subframe is 300 bits long (10 words at 30 bits each) providing 1500 (5 × 300) bit long frames. The total number of bits required to send out the total satellite information is 37.5 kbits (25 × 1500). Therefore, the time it takes to send out the total number of bits at a 50 bps rate is 12.5 minutes as shown:

$$37.5 \text{ kbits}/50 \text{ bps} = 750 \text{ sec} = 12.5 \text{ min} \qquad 9.1$$

The five subframes are specified below and contain the data necessary for GPS operation:

- Subframe 1: SV clock corrections
- Subframe 2&3: Complete SV ephemeris data. Ephemeris data contain such things as velocity, acceleration, and detailed orbit definition for each satellite.
- Subframe 4&5: Subframes 4&5 accumulated for all 25 frames provides almanac data for 1-32 satellites. The almanac data is all the position data for all the satellites in orbit.

Note that the clock corrections and space vehicle (SV) or satellite ephemeris data is updated every 30 seconds.

Each of the subframes have an ID code 001,010,011,100,101 and also uses a parity check. Parity check involves simply adding the number of "ones" sent and if an odd number is sent, the parity is a "one" and if an even number is sent, then the parity is a "zero".

Global Positioning System

9.3 GPS Receiver

The standard GPS receiver is designed to receive the C/A coded signal with a signal input of -130 dBm. Since the C/A code amplitude is 3 dB higher than P-code amplitude and the frequency of the C/A code is 1/10 of the P- code, then using a filter, the receiver is able to detect the C/A code messages. Note that the C/A code also lags the P-code by 90 degrees. For most systems the SV needs to be above 5 degrees elevation in order to keep multipath and jamming signals at a minimum. Since the GPS signal is continuous BPSK signal, a sliding correlator can be used to strip off the 1.023 MHz chipping signal leaving the 50 Hz data rate signal.

9.4 Atmospheric Errors

The atmospheric path loss for the satellite link is approximately 2 dB. The troposphere and ionosphere cause variable delays distorting TOA and position (errors are dependent on atmosphere, angle, time of day etc). These delays are extreme for low elevation satellites.
In order to compensate for the errors, the ionosphere corrections are implemented by the following using both frequencies (f_{L1}, f_{L2}):

$$PR = \frac{f_{L2}^2}{f_{L2}^2 - f_{L1}^2} (PR(f_{L2}) - PR(f_{L1}))$$

9.2

where:

> PR = compensated pseudorange
>
> f_{L2} = frequency of L2
>
> f_{L1} = frequency of L1
>
> $PR(f_{L2})$ = pseudorange of L2
>
> $PR(f_{L1})$ = pseudorange of L1

If the pseudoranges of L1 and L2 are not known, then the measured and predicted graphs over time need to be used. This would be the case for a C/A code only receiver.

The simple C/A code receiver uses the ionospheric correction data sent by the satellites for a coarse ionosphere correction solution.

9.5 Multipath Errors

The angular accuracy for platform pointing is affected by multipath as shown below:

Global Positioning System 9.3

$$\sigma_\theta = \frac{\sigma_R}{L}$$

σ_θ = *angular accurracy of platform pointing*
σ_R = *different range difference caused by multipath*
L = *Baseline*

Good antenna design can reduce multipath to approximately 1 meter. The antenna is designed to reduce ground multipath by ensuring that the gain of the antenna is low towards the ground and other potential multipath sources compared to the gain towards the satellites of interest.

Using integrated doppler (or carrier phase) can also improve the effects of multipath on the overall positioning solution to help smooth the code solution. Further discussion on smoothing the code solution with the integrated doppler of the carrier follows later in this chapter.

Another way to reduce multipath errors is to first determine that the error is caused by multipath, and then enter it in as a state variable in the Kalman filter. This solution requires that the multipath can be measured accurately and is constant.

One interesting observation is that multipath errors can actually give you a closer range measurement. By definition, multipath arrives later than the direct path, however, the resultant auto-correlation peak can be distorted such that the tracking loop (early-late gate, etc.)

334   Transceiver System Design

produces a solution that is earlier than the direct path, (see Figure 9-2).

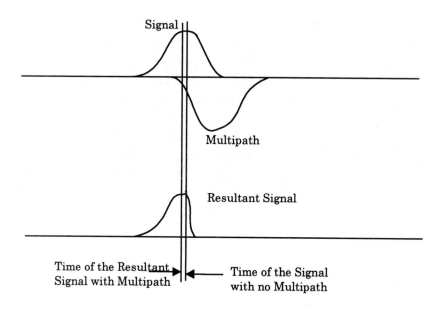

Figure 9-2 *Multipath causes resultant signal to occur earlier in time due to processing.*

9.6 Narrow Correlator

As previously stated, multipath can cause errors in range due to distortion of the auto-correlation peak in the tracking process. One way to reduce errors is to narrow the correlation peak which is known as the narrow correlator. The early-late, tau-dither loops etc., are generally operated on half-chips, that is half chip early and a half chip late which provides points on the cross-correlation peak half way down from the top on both sides. By using less than half chip dithering, the points are closer to the peak, thus reducing the ambiguity caused by multipath see Figure 9-3. The accuracy of the measurement is also greatly improved for the same reasons.

A precaution in using the narrow correlator is that the precorrelation bandwidth is larger depending on the step size of the correlator, for example:

> 2 MHz is required for 1 C/A chip($\pm$[1/2 chip]) standard correlator.
> 8 MHz is required for 0.1 C/A chip ($\pm$0.05 chip) narrow correlator.

The narrower the correlator the wider the precorrelation bandwidth needs to be. This is because the peak needs to be sharper so that the input code needs to have sharper rise times to create this sharper peak. If the peak is rounded, then the narrow correlator does not improve performance due to the ambiguity of the samples. Therefore, the drawbacks of the narrow correlator is a

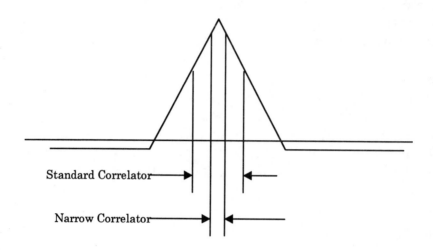

Figure 9-3 *Comparison of the standard correlator vs. the narrow correlator on the correlation peak.*

larger pre-correlation bandwidth (8 MHz compared to 2MHz) which is also more vulnerable to jammers and noise, and a faster clock is required. However, in most precision navigation applications, this technology is preferred.

For P-code receivers, the p-code null-to-null bandwidth is 20 MHz. Therefore, in order for the narrow correlator to work, the bandwidth should be approximately 80 MHz. The problem is that the satellite filters this output signal

Global Positioning System

using a bandwidth of approximately 20 MHz or so. Therefore, the increased bandwidth in the receiver does not produce sharper peaks in the autocorrelation peak because the high frequency components are filtered out at the satellite transmitter.

The locking mechanism to align the correlation peak is called a delay lock loop. The delay lock loop processing can be accomplished by searching at early and late as shown above, or by punctual and early-minus-late that generates the error off of the punctual. For low signal/noise this may prove to have an advantage since the punctual signal is higher in amplitude than the early or late gate.

9.7 Selective Availability

Selective Availability (SA) was instigated by the Department of Defense (DOD) to make the C/A code more covert. They did this by jittering or dithering the code clock and to provide an improper description of the satellite orbit, or in other words, ephemeris data corruption. This reduces the accuracy of the C/A code receiver since the method of how they are doing SA is not available to the public. Even with SA on, there is a maximum dispersion of 100 meters guaranteed at this point in time.

Note that many of the specifications written for accuracy are done with SA turned off to meet the requirements. This provides a good baseline for comparison, but caution needs to be taken when using the actual accuracy numbers since the SA is turned on most of the time.

9.8 Carrier Smoothed Code

The code solution by itself is noisy, particularly with the effects of multipath, SA, and atmosphere. This makes it difficult to obtain accurate measurements for the code derived pseudoranges and introduces variations of the measurements. To aid in the code solution, the carrier is used to help smooth the variations in the code solution.

The carrier solution is a low noise solution, however, it has a wavelength ambiguity so that it does not know which of the wavelengths that it is suppose to be comparing for the phase offset. For example, the difference in phase may be 10 degrees but may be off several cycles or wavelengths. By using the code solution which contains no wavelength ambiguity and the carrier phase solution, a carrier smoothed code can be produced. Basically, the phase change at selected repeatable points of the carrier is used to smooth the noise of the received code.

This method of measuring the carrier phase is referred to as integrated doppler and is done on each satellite. By integrating the doppler frequency of the carrier, $d\phi/dt$, ϕ is obtained. This is the phase shift that is changing with time, which is directly proportional to the change of range with time. Therefore, this phase plotted with time has the same slope as range plotted with time. The absolute range is not required since the phase measurements are used only to smooth out the code solution, the low noise carrier integrated doppler plot (corresponding to range change with time) is better than the noisy code solution of the

Global Positioning System

range which is also changing with time. The doppler is ± a frequency, and integrating gives the phase change/time. The limits of integration for the phase analysis need to be modified. Another way to evaluate the integrated doppler is to simply use the frequency times the period of integration as shown below:

$$IntegratedDoppler = \int f dt = f \times t \qquad 9.4$$

For example, if the doppler frequency is 1 kHz and the time of integration is from 0 to 0.5 seconds, then the equation becomes:

$$IntegratedDoppler = \int^{.5} 1000 dt \qquad 9.5$$
$$= 1000 \times t \ |^{.5} = 1000 \times .5 = 500 seconds$$

This is used to smooth out the code variations. For example, if the satellite is moving away from the receiver, the range (R) is positively increasing as the phase ϕ is negative increasing. If the satellite is moving towards the receiver, the range is positively decreasing while the phase is positively increasing. Note that the doppler is negative for a satellite going away so the phase ϕ is negative, and yet the range is positive. A Kalman filter is often used to incorporate the integrated doppler in the code solution. The Kalman filter is designed to achieve minimum variance and zero bias error. This means that the filter follows the mean of the time-varying signal with minimum variation to the mean. The Kalman filter uses the phase change to smooth the estimated range value of the solution as follows:

$$R_a = R_e - ID \qquad 9.6$$

where:

R_a = actual range
R_e = measured range
ID = integrated doppler

A filter could be used to smooth the noise of the code but it would be difficult to know where to set the cut off frequency to filter out the changes since this would vary with each of the satellites being tracked.

9.9 Differential GPS

Differential GPS (DGPS) uses a differencing scheme to reduce the common error in two different receivers. For example, if the GPS receivers are receiving approximately the same errors due to ionosphere, troposphere and Selective Availability, these errors can be subtracted out in the solution. This would increase the accuracy tremendously, since these errors are contributors to reduced accuracy.

Differential GPS uses a ground based station with a known position. The ground station calculates the differential pseudorange (PR) corrections, how the pseudoranges are

Global Positioning System

different from the surveyed position. These corrections are sent to other non-surveyed GPS receivers, for example an aircraft. These PR corrections are then applied to the aircraft's pseudoranges for the same satellites. The accuracy obtained by differential means can produce a position solution that is approximately 100 times more accurate than an autonomous C/A code receiver.

9.10 DGPS Time Synchronization

GPS time provided in the satellite navigation data is used to synchronize the receivers time reference. Selective Availability can affect the time reference. Differential GPS helps to reduce SA effects on both time and orbital information. The assumption is that the time stamping of the aircraft and ground station occur at the same time even though absolute time is varied by Selective Availability.

9.11 Relative GPS

Relative GPS is determination of a relative vector by one receiver, given its own satellite measurements and also the raw pseudorange measurements from the second receiver. The idea here is finding the other receiver with high accuracy such as an airplane landing on a moving platform such as an aircraft carrier. Raw pseudorange observables are uplinked to the aircraft and the aircraft calculates its position relative to the ground using both the aircraft's and ground pseudoranges. To enhance accuracy and integrity, the same set of satellites must be used.

In a relative solution, atmospheric effects are assumed to be identical for both the ground and airborne units. Therefore, no tropospheric or ionospheric corrections are performed on the independent measurements. In a relative GPS system, the absolute accuracy is dependent upon the accuracy of both the ground and airborne receivers.

9.12 Doppler

Doppler or range rate caused by the satellite moving towards or away from the GPS receiver needs to be determined so that it does not cause a problem in carrier phase tracking systems or when the carrier is used to smooth the code. Note that in a DGPS or Relative GPS system, this doppler is corrected in the process. The doppler frequency range is about ± 5 kHz for the worst case. Doppler causes the number of cycles to be off when using the carrier phase approach which is discussed later in the chapter. If doppler is known then the number of cycles are known but not λ ambiguities inherent with carrier phase tracking.

9.13 Carrier Phase Tracking

Kinematic carrier phase tracking (KCPT) is a method used for very high accurate positioning, less than 0.1 meters. KCPT is most commonly used in survey applications, but

Global Positioning System

is recently being applied to high accuracy commercial aircraft approach and landing systems. The KCPT solution uses the difference in phase of the carrier to determine range. The problem with the KCPT solution is the cycle ambiguity or wavelength ambiguity. The phase difference may be known, but the number of cycles for the range is not. Therefore, there are techniques to resolve the number of cycles.

If the receiver is moving, both the doppler caused by satellite and the doppler due to receiver movement need to be considered.

One of the concerns with the KCPT solution after acquisition is what is known as cycle slip, especially when undetected. This is where the number of calculated cycles change. If the cycle count slips, the range error is off by the number of cycles that have slipped. For L1, one cycle slip affects the pseudorange by approximately 0.6 feet or 19 cm. The system needs to detect cycle slip in order to adjust the cycle number, maintain accuracy, continuity of function, and to ensure integrity or safety of the application. There are several methods of detecting cycle slip and correcting the number of cycles. Some of these are the use of a Differential Autonomous Integrity Monitor where the Ground observables are compared to the Airborne observables for each satellite. Also, multiple solutions can be performed in parallel using the same raw measurement data, and then these multiple solutions are compared to each other to detect cycle slips. Further, other real-time tests which take into consideration bias and noise can be performed to detect cycle slips. Regardless of the method used to detect a cycle slip, the ambiguities must be resolved again using some *a priori* data from the previous resolution.

9.14 Double Difference ($\Delta\nabla$)

The double difference takes the difference between satellites and the difference between antennas, ground and air, air performing the $\Delta\nabla$. Any two receivers could be used for the $\Delta\nabla$.

The main objective is to solve for N.

$$\Delta\nabla = \Delta\phi_1 - \Delta\phi_2 = b \cdot (e_1 - e_2) + N\lambda \qquad 9.7$$

b = baseline vector
e_1, e_2 = unit vectors to satellites

Given the following absolute phase measurements below:

$$\phi = \rho + d\rho + c(dt - dT) + \lambda N - d_{ion} + d_{trop} + \varepsilon\phi \qquad 9.8$$

ρ = *geometric range*
$d\rho$ = *orbital errors*
dt = *satellite clock offset*
dT = *receiver clock offset*
d_{ion} = *ionosphere delay*
d_{trop} = *troposphere delay*
ε = *noise(rec.+multipath)*

Global Positioning System

there are two ways to achieve the double difference. One is to take the difference for one receiver and two satellites as shown below for first the carrier phase and then the code:

$$\delta\phi = \delta\rho + \delta d\rho + c(\delta(dt-dT)) + \lambda\delta N - \delta d_{ion} + \delta d_{trop} + \varepsilon\delta\phi \qquad 9.9$$

$$\delta P = \delta\rho + \delta d\rho + c(\delta(dt-dT)) + \delta d_{ion} + \delta d_{trop} + \varepsilon\delta\phi \qquad 9.10$$

The d_{ion} in the top equation is very small or zero in many cases if the ionospheric effects are the same for both the aircraft and the ground station.

The second receiver does the same and then the resultants are subtracted to achieve the double difference.

Another approach is to take the difference between the two receivers and one satellite. Then take the difference between the two receivers and the next satellite. Finally, take the difference of the two resultants.

Note that the atmospheric losses are reduced or eliminate by receiver differences. For short antenna baselines, they are eliminated. For large baselines, they are reduced inversely proportionate to the separation of the antennas.

9.15 Wide Lane/Narrow Lane

Wide Lane is used to reduce the cycle ambiguity inherent in the carrier phase tracking process. Wide Lane uses the difference in the received frequencies L1 and L2, L1–L2 which produces a lower frequency 348 MHz containing a larger wavelength, approximately 3.5 to 4.5 greater. Therefore, the ambiguity search is 3.5 to 4.5 smaller. The lower frequency contains less cycle ambiguities because there are fewer cycles over the same distance.

Narrow Lane uses the combination of L1 and L2, L1+L2, which provides more accuracy at the expense of more ambiguities to search over. A combination can be used, Wide Lane for ambiguity search and Narrow Lane for accuracy.

9.16 Summary

GPS technology is being used for many applications including; surveying, air traffic control and landing, position location for hikers, and mapping and location functions for automobiles. As this technology matures, there will be many more applications to use GPS. This technology was developed for military applications and has been adapted for commercial use using mainly the C/A code receivers. Industry is now focusing more effort in using the GPS signals-in-space in innovative ways that enhance the accuracy of the measurements and therefore improve the availability and integrity of user services.

9.17 References

[1] The Institute of Navigation, "Global Positioning System", Washington D.C., 1980.

[2] Navtech Seminars, "Fundamentals of GPS", ION GPS-93 Tutorial, 1993.

[3] Navtech Seminars, "Differential GPS", ION GPS-93 Tutorial, 1993.

[4] Navtech Seminars, "Dynamic Real Time Precise Positioning", ION GPS-93 Tutorial, 1993.

Problems

1. What is the null-to-null bandwidth of C/A code and P code GPS signals?

2. What is the theoretical process gain of C/A code and P code signals using 50 Hz data rate?

3. What are the advantages of using C/A code over P code?

4. What are the advantages of using P code over C/A code?

5. What is the advantage of using the narrow correlator detection process?

6. Name at least two disadvantages of using the narrow correlator detection process?

7. Name the two harmful effects of SA on a GPS receiver?

8. What is carrier smoothing? Why is the code more noisy than the carrier?

9. What is the main reason that DGPS is more accurate than standard GPS?

10. What is the main obstacle in providing a KCPT solution?

11. Why is the Wide Lane technique better for solving wavelength ambiguities?

10

Direction Finding and Interferometer Analysis

Direction finding is a method to determine the direction of a transmitted signal by using two antennas and by measuring the phase difference between the antennas as shown in 10-1. This process is called interferometry. In addition to using a static interferometer, further analysis needs to be done to calculate the direction when the interferometer baseline is dynamic. That is, the interferometer is moving and rotating in a three dimensional plane. Therefore, coordinate conversion processes need to be applied to the non-stabilized antenna baseline to provide accurate measurement of the direction in a three dimensional plane.

10.1 Interferometer Analysis

For a non-stabilized antenna baseline, the elevation, roll, and pitch of the antenna baseline need to be included as part of the interferometer process for accurate azimuth determination using interferometer techniques. Therefore, the azimuth angle calculation involves a 3-dimensional

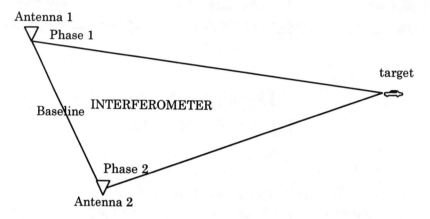

Figure 10-1 *Basic interferometer used for direction finding.*

solution using coordinate conversions and direction cosines. The overall concept is that the coordinates are moved due to the movement of the structure. Therefore the direction phasor is calculated on the moved coordinates and then a coordinate conversion is done to put the phasor on the absolute coordinates. This will become apparent in the following discussions. A brief discussion on direction cosines follows since they are the basis for the final solution.

10.2 Direction Cosines

Direction cosines provide a means of defining the direction

Direction Finding and Interferometer Analysis

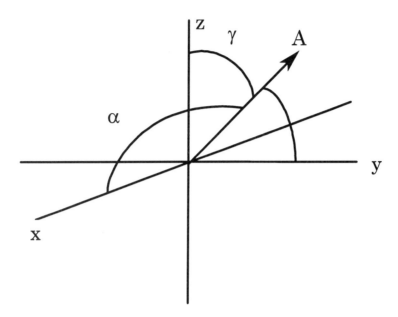

Figure 10-2 *Angles for computing the direction cosines.*

for a given phasor in a three dimensional plane. They are the key in defining and calculating the interferometer equations.

A phasor A is represented as:

$$A = |A|(\cos\alpha\, i + \cos\beta\, j + \cos\gamma\, k) \qquad 10.1$$

The i,j,k are unit vectors. For example, i = (1,0,0), j =

(0,1,0), and k = (0,0,1) related to the magnitudes in the x,y,z respectively. For example, the i is just for the x direction. Therefore, α is the angle between phasor A and the x axis, β is the angle between the phasor A and the y axis, and γ is the angle between the phasor A and the z axis as shown in Figure 10-2.

The direction cosines are the cosines of each of the angles specified, $\cos\alpha$, $\cos\beta$, and $\cos\gamma$. The cosines are equal to the adjacent side which is the projection on the specified axis divided by the magnitude of phasor A. The projection on the x axis for example is simply the x component of A. It is sometimes written as $comp_b A$. The vector A can be described by (x,y,z). The direction cosines are defined in the same manner, with x,y,z replacing i,j,k.

Another phasor is defined and is the horizontal baseline in the earth's plane. This is the baseline vector between two interferometer antennas. Therefore:

$$B = |B|(\cos\alpha x + \cos\beta y + \cos\gamma z) \qquad 10.2$$

If $\gamma = 90$, which means that the phasor is horizontal with no vertical component, then

$$A = |A|(\cos\alpha x + \cos\beta y + 0) \qquad 10.3$$

If the baseline is on the y axis, then

$$B = |B|(\cos 90 x + \cos 0 y + \cos 90 z) = |B|(0x + 1y + 0z) \qquad 10.4$$

Therefore:

Direction Finding and Interferometer Analysis 355

$$A \bullet B = |A| |B| \cos\beta \qquad 10.5$$

For a two-dimensional case, this dot product $= |A| |B| \sin\alpha$ as shown in Figure 10-3.

For this example, if the baseline interferometer is mounted on the y axis, then α is the angle from boresight, or the x axis, to the baseline which is the desired angle. This leads to the standard interferometer two dimensional equation.

10.3 Basic Interferometer Equation

The basic interferometer equation, used by many for approximate calculations, is only a two dimensional solution and is shown below:

$$dp = (2\pi d)/\lambda \ \sin\theta \qquad 10.6$$

where:
- dp = measured electrical phase difference in radians.
- d = separation of the interferometer antennas
- θ = true azimuth angle

This is a familiar equation used in most text books and in direction finding (DF) literature. This equation is derived

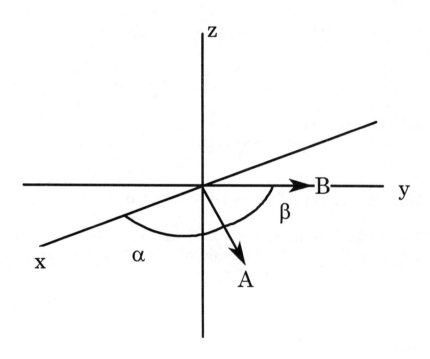

Figure 10-3 *Interferometer analysis with the baseline on the Y axis.*

by simply taking the dot product of the direction phasor with the interferometer baseline which produces the cosine of the angle from the baseline. Using simple geometry, the relationships between the additional distance traveled for one interferometer element (d2) with respect to the baseline difference (d1) and the measured electrical phase

Direction Finding and Interferometer Analysis 357

difference (dp) are easily calculated and the resulting equations are shown below (see Figure 10-4):

$$d2 = d1 \cos(\phi) \qquad 10.7$$

ϕ = angle of the phasor from the baseline

$$dp = (d2/\lambda)2\pi \qquad 10.8$$

Since the angle is usually specified from the boresight, which is perpendicular to the interferometer baseline, the equation uses the sine of the angle between the boresight and the phasor (θ) which results in the standard interferometer equation as shown below (see Figure 10-4):

$$dp = (2\pi d1)/\lambda \sin\theta \qquad 10.9$$

This still assumes that the target is a long distance from the interferometer with respect to the distance between the interferometer elements, which is a good assumption in most cases.

This works fine for a 2-dimensional analysis and gives an accurate azimuth angle with slight error for close in targets since $d\sin(\theta)$ is a geometrical estimate and A1 and A2 are assumed parallel as shown in Figure 10-4.

358   Transceiver System Design

Two-Dimensional Interferometer Analysis

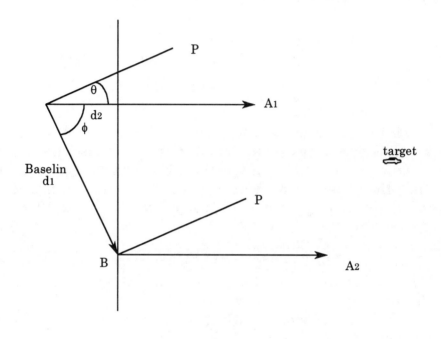

$A \cdot B = ab\cos(\phi)$
$d_2/d_1 = \cos(\phi)$
$d_2 = d_1 \cos(\phi)$
$dp = (d_2/\lambda) * 2\pi$ radians
$dp = (2\pi/\lambda) * d_2$ radians
$dp = (2\pi/\lambda) * d_1 \cos(\phi)$ radians
$\cos(\phi) = \sin(\theta)$

Therefore:
$\mathbf{dp = (\pi d_1/\lambda)\ \sin(\theta)\ radians}$

Standard 2-dimensional interferometer equation.

Figure 10-4 *Two dimensional interferometry.*

10.4 Three Dimensional Approach

If γ is not equal to 90 degrees, then

$$A = |A|(\cos\alpha\, x + \cos\beta\, y + \cos\gamma\, z) \qquad 10.10$$

$$B = |B|(0x + 1y + 0z) \qquad 10.11$$

The dot product remains the same, $A \cdot B = |A||B|\cos\beta$ however, both the angles α and β change and the dot product no longer equals $|A||B|\sin\alpha$. For example if γ is 0, α and β would be 90 degrees. Therefore, for a given A with constant amplitude, the α and β change with γ.

If the alignment is off or there is pitch and roll, these angles change. Since the angle offsets are from the mounted baseline, then B vector is defined with the angles offset from the B baseline as shown in Figure 10-5.

Therefore the resultant equation is:

$$B = |B|(\sin\alpha_1 x + \cos\beta_1 y + \sin\gamma_1 z) \qquad 10.12$$

where:

$\alpha_1, \beta_1, \gamma_1$ = the angles from the B desired baseline caused by misalignment or movement.

Therefore, if the interferometer baseline is rotated, pitched and rolled, then the results are:

$$A = |A|(\cos\alpha\, x + \cos\beta\, y + \cos\gamma\, z) \qquad 10.13$$

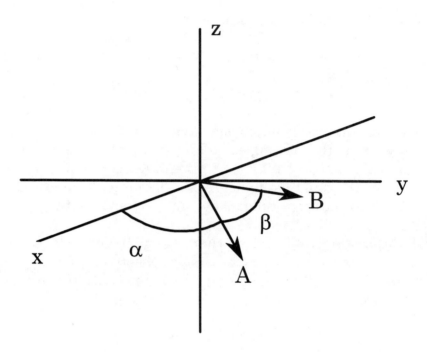

Figure 10-5 *Interferometer with the baseline vector not on the Y axis.*

$$B = |B|(\sin\alpha_1 x + \cos\beta_1 y + \cos\gamma_1 z) \qquad 10.14$$

This B is the new phasor, offset from the earth's coordinate system.

Direction Finding and Interferometer Analysis   361

$A \cdot B = |A| |B| (\cos\alpha\sin\alpha_1 x + \cos\beta\cos\beta_1 y + \cos\gamma\sin\gamma_1 z)$ 10.15

The main solution to this problem is to find the angle from boresight which is a three dimension problem when considering the elevation angle and the dynamics of the baseline.

Approaching the problem in 3-dimensions using direction cosines produces a very straight forward solution and the results are shown in the following analysis. The standard interferometer equation is not used and $\sin(\theta)$ is meaningless for this analysis since a two-dimensional dot product to resolve azimuth is not used. The effects that the elevation angle has on the azimuth solution is shown in Appendix C.

10.5 Antenna Position Matrix

The first part of the analysis is to define the three dimensional interferometer coordinate system relative to the dynamic coordinate system. The definition of this axis is imperative for any analysis. For example, mounting an interferometer on a ship produces the following analysis using the y coordinate from port to starboard (starboard is positive y) with the interferometer baseline along the y axis, the x coordinate from bow to stern (bow is positive x), and the z coordinate up and down (up is positive z) see Figure 10-6. The position matrix for the interferometer on the ship's coordinate system is defined as $\{\sin(\alpha), \cos(\beta), \sin(\gamma)\}$.

362   Transceiver System Design

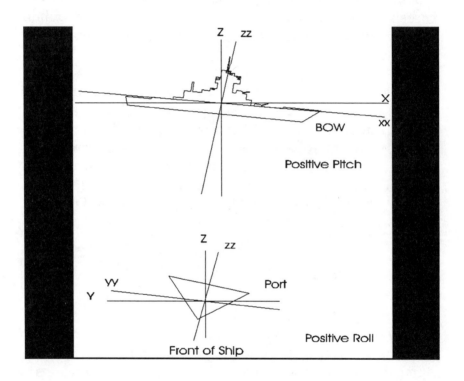

Figure 10-6 *Roll and pitch definitions.*

Direction Finding and Interferometer Analysis

If there is a misalignment in the x-y plane and not in the z plane, the position matrix is $\{\sin(\alpha),\cos(\beta),0\}$ as shown in Figure 10-7. If there are no offsets, then the position matrix is (0,1,0), which means that the interferometer is mounted along the y axis. The antenna position matrix compensates for the misalignment of the interferometer

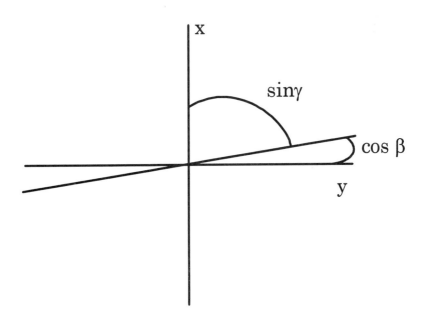

Figure 10-7 *Position matrix offset.*

364   Transceiver System Design

baseline with respect to the ship's coordinate system.

10.6   Coordinate Conversion due to Pitch and Roll

The coordinate conversion transformations modify the antenna position matrix to obtain the earth coordinate antenna position x,y,z. This is done for Yaw*, Pitch, and Roll. Heave is generally insignificant but can be included for specific cases. Yaw is the movement in the horizontal plane that affects bearing with positive Yaw rotating in the clockwise direction. Pitch is the bow of the ship moving up and down with down being in the positive direction. Roll is the port and starboard parts of the ship rotating up and down with port side down being positive. Heave is the movement of the ship as an entire unit in the vertical direction with positive heave being up. The conversion matrices are shown below. This example is done for 15 degrees roll, 5 degrees pitch, and 3 degrees yaw. The order of the analysis is yaw, pitch and roll. The order of the transformation needs to be specified and different orders will alter the final solution. Further information on the coordinate transformations are included in Appendix A. The interferometer measurement is done with the ship's antenna position at the location caused by the yaw, pitch and roll. To bring the coordinates back to earth or level coordinates, the process is done in the reverse order and opposite direction. That is, − roll, − pitch, − yaw*. The movement is defined as positive pitch is bow down and positive roll is port down. The negative numbers are put in the original coordinate transformations. This is shown in the example below.

Direction Finding and Interferometer Analysis

Roll

$$\begin{matrix} 1 & 0 & 0 \\ 0 & \cos(r) & \sin(r) \\ 0 & -\sin(r) & \cos(r) \end{matrix}$$

Pitch

$$\begin{matrix} \cos(p) & 0 & -\sin(p) \\ 0 & 1 & 0 \\ \sin(p) & 0 & \cos(p) \end{matrix}$$

Yaw

$$\begin{matrix} \cos(y) & \sin(y) & 0 \\ -\sin(y) & \cos(y) & 0 \\ 0 & 0 & 1 \end{matrix}$$

10.16

*Note: The Yaw can be solved after taking the Roll and Pitch azimuth calculation and simply offsetting the angle by the amount of the Yaw. This may prove simpler for implementation.

10.7 Using Direction Cosines

Now that the coordinate conversion and position matrix have been solved, the x,y,z values as a result of the these processes are used with the direction cosines to achieve the solution. The basic direction cosine equation as mentioned before is:

$$a = |a|(\cos\alpha x + \cos\beta y + \cos\gamma z) \qquad 10.17$$

For a unit vector, the scalar components are the direction cosines:

$$u = \cos\alpha x + \cos\beta y + \cos\gamma z = Xx + Yy + Zz \qquad 10.18$$

The x,y,z are the coordinate converted and offset compensated x,y,z. The X,Y,Z are the direction cosines of the Poynting vector for the horizontal baseline coordinate system. The unit vector is equal to the phase interferometer difference divided by the phase gain.

Since x,y,z are known from previous analysis, and u is measured and calculated, the direction cosines are the only unknowns and are solved by the following identities and equations:

$$(X^2 + Y^2 + Z^2)^{1/2} = 1 \text{ so } X^2 + Y^2 + Z^2 = 1 \qquad 10.19$$

Note that $Z = \cos\gamma = \sin\Psi$, since the direction cosine is defined from the top down and Ψ = angle from the horizontal up. The elevation angle Ψ, is calculated by using altitude and range.

Direction Finding and Interferometer Analysis

Also, the effects of the earth's surface can cause the elevation angle to have some slight amount of error. This can be compensated for in the calculation of the elevation angle and is shown in Appendix D.

Therefore:

$$X^2 + Y^2 + \sin^2\Psi = 1 \qquad 10.20$$

and from above:

$$u = Xx + Yy + \sin\Psi z \qquad 10.21$$

Solving for Y:

$$Y = (u - Xx - \sin\Psi z)/y \qquad 10.22$$

Solving for X^2:

$$X^2 = 1 - \sin^2\Psi - [(u - Xx - \sin\Psi z)/y]^2 \qquad 10.23$$

Therefore, putting the equation in quadratic form (the rest of the math is left for the reader). To do this in steps is much easier. Solving simultaneous equations, produces a quadratic equation for X in the form $AX^2 + BX + C$ where:

$$[1+(x/y)^2]X^2 - [2x/y(u - \sin\Psi z)/y]X + \sin^2\Psi - 1 + [(u - \sin\Psi z)/y]^2 = 0 \qquad 10.24$$

where:

$$A = [1 + (x/y)^2] \qquad 10.25$$

$$B = [2x/y(u - \sin\Psi z)/y]$$
$$= [-2x/y^2(u - z\sin(\Psi))] \qquad 10.26$$

u = Electrical phase difference in radians (phase interferometer difference divided by the phase gain).

$$C = \sin^2\Psi - 1 + [(u - z\sin\Psi)/y]^2 \qquad 10.27$$

Therefore, solving for X using the quadratic equation:

$$X = \frac{-B + \sqrt{B^2 - 4AC}}{2A} \qquad 10.28$$

Solving for Y:

$$Y = \frac{u - xX - z\sin\psi}{y} \qquad 10.29$$

Then the azimuth angle θ is:

$$\theta = \operatorname{atan}(Y/X) \qquad 10.30$$

Direction Finding and Interferometer Analysis   369

The azimuth angle using the previous analysis is shown in Figure 10-8.

The direction cosines are the values in the x and y directions in the final converted coordinate system, Xx + Yy + Zz. The x,y,z coordinates include the coordinate conversion constants.

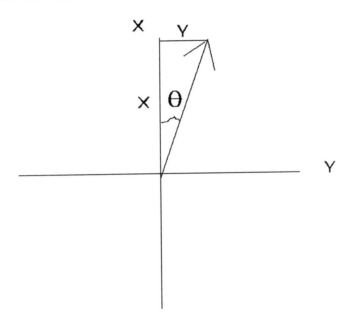

Azimuth = ATan(Y/X)

Figure 10-8 *Baseline located on the Y axis.*

Transceiver System Design

Table 10-1 *Interferometer analysis example.*

EXAMPLE: 15,5,3 DEGREES ROLL,PITCH,YAW, 45 DEGREES EL AT 24 DEGREES AZ									
x-axis on the bow-stern axis									
y-axis on port-starboard axis			Roll=	-15		Roll and Pitch Only:			
						A	B	C	
			Roll Matrix		Pos. Mat.	1.0005454	-0.012517	-0.42818	
x`=	0	x`=	1	0	0	0 x``			
y`=	0.9659258	y`=	0	0.9659258	-0.258819	1 y``	X	Y	Az. Angle
z`=	0.258819	z`=	0	0.258819	0.9659258	0 z``	0.6604613	0.2525685	20.927463
									ATAN(Y/X)
			Pitch =	-5					
			Pitch Matrix			Roll, Pitch, Yaw:			
x =	0.0225576	x =	0.9961947	0	0.0871557	0 x`			
y =	0.9659258	y =	0	1	0	0.9659258 y`	A	B	C
z =	0.2578342	z =	-0.087156	0	0.9961947	0.258819 z`	1.0008421	0.015556	-0.428159
			Yaw =	-3			X	Y	Az. Angle
							0.6463378	0.2867882	23.927463
			Yaw Matrix						
	-0.028026	xyaw =	0.9986295	-0.052336	0	0.0225576 x	Note that the Yaw is simply		
	0.9657826	yyaw=	0.052336	0.9986295	0	0.9659258 y	an offset of 3 degrees.		
	0.2578342	zyaw=	0	0	1	0.2578342 z			
			Radians	Azimuth:	Answer:	Error:			
Elevation Angle (deg.) =			45	0.7853982	24	23.927463	0.0725368		
Electrical Phase Diff.(deg) =			635.295	25.277585	Elect.Phase/Phase Gain				
Electrical Phase Diff.(rad) =			11.087989	38.602861					
Phase Gain =			25.132741		0.4411771	Elec.Phase Diff.(rad)/phase gain			
For an error in alignment, so that the interferometer is not lined up with the axis,									
the following example is shown:									
Alignment Error:		Inputs:	Calculations:						
	Y =	0	Angle =	0					
	Dist.	1.312	x`` =	1					
			y`` =	0					

Direction Finding and Interferometer Analysis 371

An example showing the steps in producing the interferometer solution is found in Table 10-1. This is programmed into a spread sheet for ease in changing parameters for different interferometer configurations. Other factors that are considered when determining azimuth are true north calculations and phase ambiguities which are included in Appendix B.

10.8 Alternate Method

The above analysis was done with the X axis on the bow-stern axis and the Y axis on the port-starboard axis. If another axis is used, then the equations need to be modified to reflect the correct axis. If the X axis is on the port-starboard axis and the Y axis on the bow-stern, then a right hand coordinate system is defined. This analysis can be used with the following changes:

    a.   The quadratic equation for solving x needs to be:

$$X = \frac{-B - \sqrt{B^2 - 4AC}}{2A} \qquad 10.31$$

    b.   The azimuth angle is calculated as:

$$\phi = \text{atan}(X/Y) \qquad 10.32$$

Notes: The formula for converting the differential phase in electrical degrees to angular degrees is:

$$\text{Angular\_degrees} = (1/\text{phase\_gain}) * \text{differential\_phase} \qquad 10.33$$

where:

>angular_degrees = the number of degrees in azimuth that the target is from the boresight and right of boresight is defined as positive.

>phase_gain = $(2\pi d)/$wavelength.

>differential_phase = the difference in phase between the two antennas. This parameter is measured in electrical degrees.

10.9 Summary

Interferometers use phase differencing to calculate the direction of the source of transmission. The basic mathematical tool used in interferometer calculations is the direction cosine. Direction cosines define the direction of the phasor in a three-dimensional plane. The basic interferometer equation only deals with two dimensional analysis. The third dimension alters the two-dimensional solution significantly. This three-dimensional approach using coordinate conversion techniques to compensate for baseline rotations provides an accurate solution for the interferometer.

10.10 References

[1] Saturnino L. Salas, Einar Hille, *Calculus*, John Wiley & Sons, New York, 1974.

[2] David Halliday, Robert Resnick, *Fundamental of Physics*, John Wiley & Sons, 1974.

[3] Don Shea, "Notes on Interferometer Analysis", 1993.

Problems

1. What is the true azimuth angle for a two dimensional interferometer given that the operational frequency is 1 GHz, the phase different between the two antennas is 10 radians and the antennas are separated by 3 meters?

2. Why is a three dimensional approach required for a typical interferometer calculation?

3. Why is the sin θ used for the elevation angle calculation?

4. Is the order of the coordinate conversion process critical? Why?

Appendix

Appendix A

The following are diagrams showing the coordinate conversions for motion including Heading, Roll, Pitch and Yaw.

SHIP'S HEADING

(AFFECTS THE ROTATION IN THE X AND Y PLANE)

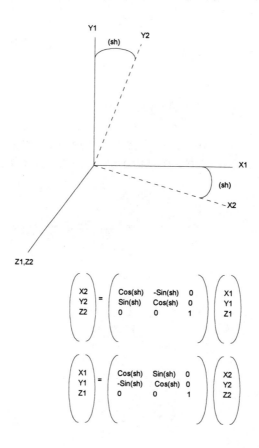

$$\begin{pmatrix} X2 \\ Y2 \\ Z2 \end{pmatrix} = \begin{pmatrix} \cos(sh) & -\sin(sh) & 0 \\ \sin(sh) & \cos(sh) & 0 \\ 0 & 0 & 1 \end{pmatrix} \begin{pmatrix} X1 \\ Y1 \\ Z1 \end{pmatrix}$$

$$\begin{pmatrix} X1 \\ Y1 \\ Z1 \end{pmatrix} = \begin{pmatrix} \cos(sh) & \sin(sh) & 0 \\ -\sin(sh) & \cos(sh) & 0 \\ 0 & 0 & 1 \end{pmatrix} \begin{pmatrix} X2 \\ Y2 \\ Z2 \end{pmatrix}$$

Direction Finding and Interferometer Analysis for a Non-stabilized Antenna Baseline

SHIP'S ROLL

(AFFECTS THE ROTATION IN THE X AND Z PLANE)

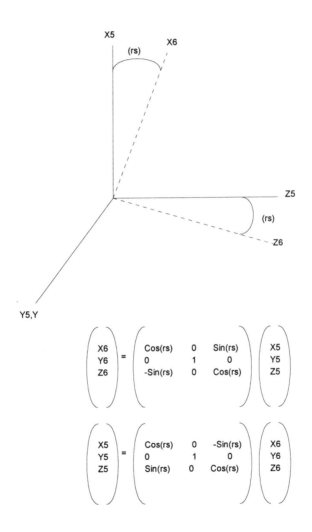

$$\begin{pmatrix} X6 \\ Y6 \\ Z6 \end{pmatrix} = \begin{pmatrix} \cos(rs) & 0 & \sin(rs) \\ 0 & 1 & 0 \\ -\sin(rs) & 0 & \cos(rs) \end{pmatrix} \begin{pmatrix} X5 \\ Y5 \\ Z5 \end{pmatrix}$$

$$\begin{pmatrix} X5 \\ Y5 \\ Z5 \end{pmatrix} = \begin{pmatrix} \cos(rs) & 0 & -\sin(rs) \\ 0 & 1 & 0 \\ \sin(rs) & 0 & \cos(rs) \end{pmatrix} \begin{pmatrix} X6 \\ Y6 \\ Z6 \end{pmatrix}$$

SHIP'S PITCH

(AFFECTS THE ROTATION IN THE Y AND Z PLANE

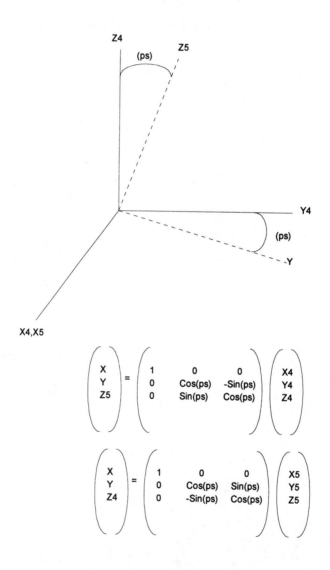

$$\begin{pmatrix} X \\ Y \\ Z5 \end{pmatrix} = \begin{pmatrix} 1 & 0 & 0 \\ 0 & \cos(ps) & -\sin(ps) \\ 0 & \sin(ps) & \cos(ps) \end{pmatrix} \begin{pmatrix} X4 \\ Y4 \\ Z4 \end{pmatrix}$$

$$\begin{pmatrix} X \\ Y \\ Z4 \end{pmatrix} = \begin{pmatrix} 1 & 0 & 0 \\ 0 & \cos(ps) & \sin(ps) \\ 0 & -\sin(ps) & \cos(ps) \end{pmatrix} \begin{pmatrix} X5 \\ Y5 \\ Z5 \end{pmatrix}$$

Direction Finding and Interferometer Analysis for a Non-stabilized Antenna Baseline

SHIP'S YAW

(AFFECTS THE ROTATION IN THE X AND Y PLANE, SAME AS HEADING)

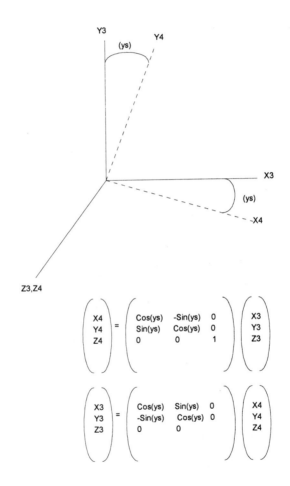

$$\begin{pmatrix} X4 \\ Y4 \\ Z4 \end{pmatrix} = \begin{pmatrix} \cos(ys) & -\sin(ys) & 0 \\ \sin(ys) & \cos(ys) & 0 \\ 0 & 0 & 1 \end{pmatrix} \begin{pmatrix} X3 \\ Y3 \\ Z3 \end{pmatrix}$$

$$\begin{pmatrix} X3 \\ Y3 \\ Z3 \end{pmatrix} = \begin{pmatrix} \cos(ys) & \sin(ys) & 0 \\ -\sin(ys) & \cos(ys) & 0 \\ 0 & 0 & 0 \end{pmatrix} \begin{pmatrix} X4 \\ Y4 \\ Z4 \end{pmatrix}$$

Appendix B

True North Calculations:

The formula to be used for correcting for true north is:

bearing_true(n) = Relative_bearing(n) + ships_heading(n)

where:

bearing_true(n) = The target bearing from the ship with respect to true north.

ships_heading = The ship's heading as measured by the Ship's Inertial Navigation System or a magnetic compass.

Relative_bearing(n) = The current target bearing relative to ship's longitudinal axis starting at the bow for zero degrees relative and rotating clockwise.

The formula to be used for correcting from true north to magnetic north is:

bearing_mag(n) = bearing_true(n) + magnetic_variation

where:

bearing_mag(n) = The target bearing from the ship with respect to magnetic north

   bearing_true(n) = The target bearing from the ship

Direction Finding and Interferometer Analysis for a Non-stabilized Antenna Baseline

with respect to true north.

magnetic_variation= the variation between magnitic north and true north. Magnetic variation may be plus or minus.

Phase Ambiguities:

For an interferometer to have no phase ambiguities, the spacing between the antennas should be less that $\lambda/(2\pi)$ wavelength apart. This provides a phase gain of less than 1. Note that a phase gain of exactly 1 gives a one-to-one conversion from azimuth angle to electrical angle. Therefore, there are no phase ambiguities. If the separation is greater than 1 wavelength, giving a greater than 1 phase gain, then ambiguities exist. For example, if the separation is 2 λ, then half the circle covers 360 degrees and then repeats for the second half of the circle. Therefore, a phase measurement of 40 degrees could be two spacial positions.

Appendix C

Elevation effects on azimuth error:

The elevation effects on the azimuth error are geometric in nature and were evaluated to determine if they are needed in the azimuth determination and to calculate the magnitude and RMS azimuth error. A simulation was done using three different angles 10, 25, and 45 degrees. The simulation sweeps from 0 to 24 degrees in azimuth angle and the error is plotted in degrees. The results of the simulations are shown in Figure C-1.

The azimuth error is directly proportional to the elevation angle, the higher the angle the greater the error. Also, the azimuth error is directly proportional to the azimuth angle off interferometer boresight. This analysis was performed with a horizontal interferometer baseline with no pitch and roll. The azimuth error at an azimuth angle of 22.5 degrees and an elevation angle of 45 degrees was equal to 6.8 degrees which is the worst case error.

The azimuth error caused by elevation angle can be calculated by:

Az Error = True Az
− asin[cos(true el)sin(true az)/cos(assumed El angle)]
= 22.5 − asin(cos(45)sin(22.5)/cos(0)) = 6.8 degrees

For elevation compensation only, an approximate solution can be used and is shown below:

Direction Finding and Interferometer Analysis for a Non-stabilized Antenna Baseline

$$dp = (2\pi d)/\lambda \ \sin\theta \ \cos\Psi$$

where:

Ψ = elevation angle

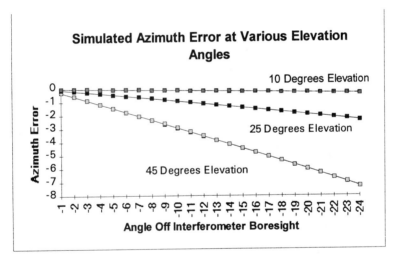

Figure C-1 *Azimuth error due to elevation angle.*

However, including roll and pitch produces a 3-dimensional analysis.

Using the standard interferometer, 2-dimensional equation, and trying to compensate for the elevation, roll, and pitch, results in an very complex transcendental equation, without making too many assumptions, and is generally solved by iterative methods.

Direction Finding and Interferometer Analysis for 385
a Non-stabilized Antenna Baseline

Appendix D

Earth's Radius Compensation for Elevation Angle Calculation

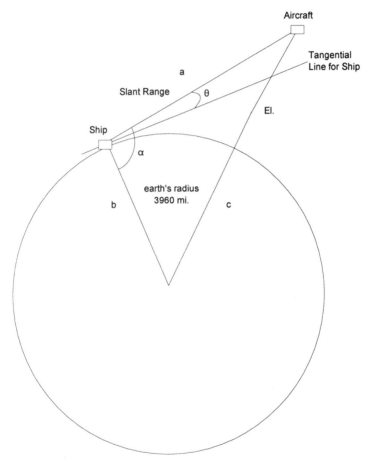

Figure D-1 *Earth's radius compensation for elevation angle calculation.*

The angle θ, which is the desired elevation angle, is calculated for the curved earth as:

1. Solve for the angle alpha using law of cosines:

    $$c^2 = a^2 + b^2 - 2ab\cos(\alpha)$$

    $$\alpha = \mathrm{acos}((a^2 + b^2 - c^2)/2ab)$$

where:

    a = slant range

    b = altitude of the ship plus the earth's radius

    c = altitude of the aircraft plus the earth's radius

Therefore:

$$\theta = \alpha - 90\ \mathrm{deg.} = \mathrm{acos}((a^2 + b^2 - c^2)/2ab) - 90\ \mathrm{deg.}$$

Index

A
A/D 98
Adaptive filter 279
Adaptive line enhancer 282
AGC 140
    amplifier curve 142
    detector 144
    linearizer 144
    loop filter 149
    modulation frequency distortion 156
    phase-lag filter 151
    threshold level 151
Antenna gain 12
Autocorrelation 190
Automatic frequency control 197

B
BER 177, 228
Binomial Distribution Function 235
Bit synchronizer 199
BPSK 191
Burst clamp 273
Burst jammer 272

C

C/A code 328
Carrier recovery 198
Carrier smoothing 338
Channelized receiver 319
Chirped-FM 82
Coding gain 41
Coherent demodulation 191
Coordinate conversion 364
Correlator 179
Costas loop 191, 195
Crystal video receiver 316
Cumulative distribution function 219, 233

D

Damping ratio 154
Data-aided loop 198
Delay-lock loop 190
Despreading correlator 187
Differential GPS 340
Digital signal processing 202
Diplexer 52
Direction cosine 366
Direction cosines 352
Dispersion 202
Doppler 342
Double difference 344
DPSK 229
Duplexer 52
Dynamic range 95
    1 dB compression point 110
       Amplitude 100

Index 389

   Example 110
   Frequency 101
   Single Tone 102
   System 109
   Tangential sensitivity 113
   Two tone 102
   With AGC 111

E
Early-late gate 189
E_b/N_o 229
Effective isotropic radiated power 14
Electronic counter counter measures 204
Electronic counter measures 204
Erf function 222
Erfc 233
Eye pattern 199

F
Fading 247
FIR 179, 281
FIR filter 182
Frequency hopping 79
Frequency shift keying 78
   MSK 78
Fresnel zone 248
FSK 229

G
Gaussian process 220
Glint 247
GPS

　　　　Narrow Lane 346
　　　　Wide Lane 346
GPS receiver 331
Gram-Schmidt Orthogonalizer 305

I
Image rejection 95
Instantaneous Fourier transform 317
Instantaneous frequency measurement 316
Integrated doppler 339
Intercept receivers 314
Interferometer 353
Intermods 104
Intersymbol interference 198, 203

K
Kalman filter 339
KCPT 342

L
Limiter 94
Link budget 3
　　　　Link margin 42
LMS 285
Loss
　　　　Free space 16
　　　　Propagation 17

M
Matched filter 178, 197
Maximally length sequence 86
Mean-squared error 286

Index 391

Microscan Receiver 318
Minimum discernable signal 11, 91, 95
Mixer
    Spur analysis 124
    Image reject 124
MSK 229
Multipath 247, 252
    Diffuse 248, 253
    Divergence factor 256
    Grazing angle 249
    Scattering coefficient 251
    Smooth reflection 249
    Specular 248
Multiple Users
    CDMA 83
    FDMA 83
    TDMA 83

N
Narrow correlator 335
Natural frequency 154
Noise
    Factor 37, 115
    Figure 35
    kTB 120
    kTBF 35
    Phase noise 120
Noise factor 37
Noise figure 35

P

P-code 328
Phase shift keying
    16OQPSK 75
    BPSK 62
    D8PSK 73
    DPSK 64
    Higher order 69
    MSK 76
    Oqpsk 68
    $\pi/4$ 72
    QPSK 65
PLL 159, 196
PN code 188, 240
PN code generator 86
POE 228
Power
    dB 5
    dBm 9
    EIRP 14
Power amplifier
    Classes 59
Power spectral density 219
PPM 178, 183
Pre-aliasing 130, 237
Probability density function 214
Probability of detect 232
Probability of error 41
Probability of false alarm 231
Propagation losses 17

Index

Q
QPSK 194, 229
Quadrature
    Downconversion 56
    Upconversion 56
Quantization error 132, 223

R
Relative GPS 341

S
Selective Availability 337
Shape factors 129
Signal to noise 40
    E_b/N_o 40
Squaring loop 191, 192
Standard deviation 218
Superheterodyne 92

T
Tau-dither loop 189
TDMA 181
Time Hopping 82
TOA 178

V
Variance 217
VSWR 60

Answers

Chapter 1:

1. Answer: 0 dBm is a power level of 1 milliwatt. ± 2 dBm are two power levels of 1.58 milliwatts and .63 milliwatts. 0 dBm − 2 dBm = 1 mW − 1.58 mW = −.58 mW. Cannot have negative power. The correct answer would be 0 dBm ± 2 dB = ±2 dBm = 1.58 mW and .63 mW.

2. Answer: 0.00025 mW. −36 dBm. $10\log(.00025) = -36$ dBm

3. Answer: $10\log 1 = 0$ dBm. $10\log(.01) = -20$ dBm. Therefore, the spurious response is −20 dBc. For −40 dBc, the spur level is 0 dBm − 40 dBc = −40 dBm = .1 µW.

4. Answer. The MDS will be reduced by 1.5 dB. Increase the transmitter power by 1.5 dB, reduce the losses before the LNA by 1.5 dB, or reduce the noise figure of the LNA by 1.5 dB.

5. Answer. 1.91 meters

6. Answer. −131.8 dB of attenuation. $A_{fs} = 20\log[c/4\pi Rf]$

7. Answer: −174 dBm + 10log(10MHz) + 3 dB = −107 dBm.

8. Answer. You must convert to actual power and use the noise factor equation and then convert the answer to dB. The results are: F = 1.995 + [(10 + 3.16 −1)/10] = 3.211: 10log(3.211) = 5.07 dB.

9. Answer: 2 × chip rate = 40 MHz null-to-null bandwidth.

10. Answer: −174 dBm + 10log(10MHz) + 5 dB + 3 dB + 25 dB + 15 dB = −59 dBm Output S/N (dB) = −36 dBm − (−59 dBm) = 13 dB S/N.

Answers

Chapter 2

1. The transmitted signal is:

$$\cos(.1MHz)t\cos(10MHz)t$$
$$=\frac{1}{2}\cos(10.1MHz)t + \frac{1}{2}\cos(9.9MHz)t$$

The received signal using the worst case by multiplying by sin(10MHz)t is:

$$[\frac{1}{2}\cos(10.1MHz)t + \frac{1}{2}\cos(-9.9MHz)t][\sin(10MHz)t]$$

$$=\frac{1}{4}[\sin(20.1MHz)t + \sin(.1MHz)t]$$
$$-\frac{1}{4}[\sin(.1MHz)t + \sin(-19.9MHz)t]$$
$$=\frac{1}{4}[\sin(20.1MHz)t + \sin(.1MHz)t]$$
$$-\frac{1}{4}[\sin(.1MHz)t + \sin(-19.9MHz)t]$$
$$=\frac{1}{4}[\sin(20.1MHz)t - \sin(-19.9MHz)t]$$

The signal frequency is cancelled out.

2. The transmitted filtered waveform is as follows:

$$\frac{1}{2}\cos(10.1 MHz)t + \frac{1}{2}\cos(9.9 MHz)t$$

$$= \frac{1}{2}\cos(9.9 MHz)t \qquad \textit{for filtered output.}$$

The received signal using the worst case by multiplying by sin(10MHz)t is:

$$\frac{1}{2}\cos(9.9 MHz)t][\sin(10 MHz)t]$$

$$= \frac{1}{4}\sin(19.9 MHz)t - \frac{1}{4}\sin(.1 MHz)t$$

By filtering the signal waveform, sin(.1 MHz) is retrieved.

Answers   399

3. The phasors are shown for the summation of two QPSK modulators to produce 8-PSK generator. The possible phases if the phasor diagram is rotated by 22.5 degrees are: 0, 45, 90, 135, 180, 225, 270, and 315. The phase between the two QPSK modulators is 45 degrees.

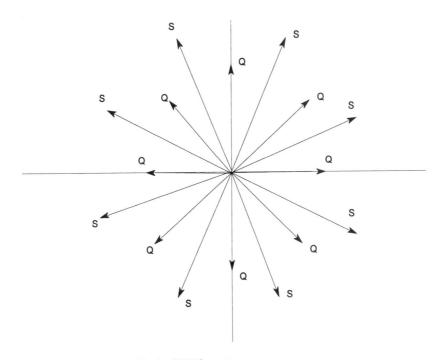

Q1 = the phasors of the first QPSK Generator

Q2 = the phasors of the second QPSK Generator offset in phase from the first by 45 degrees.

S8 = the 8 resultant phasors to make up the 8-PSK modulator

4. The possible phase shifts using two OQPSK modulators can be determined by the phasor diagram and are ± 45, ± 90, ± 180.

5. By building two QPSK modulators, and then off-setting the modulators in time so that only one QPSK modulator switches phase at one given time. This prevents the 180 degree phase shift. The 180 degree phase shift is a problem because it produces unwanted AM on the waveform since the resultant phasor travels through 0 amplitude on the transition. Since a practical system cannot process a instantaneous signal, (infinite bandwidth), then this AM is present.

6. The "1" is necessary since the output of the modulo-2 adder will be "0" and never change the pattern in the shift registers, all zeros is the resultant and never changes.

7. The process gain is equal to:

$$10\log(1/\text{duty cycle}) = 10\log(1/.2) = 6.99 \text{ dB}$$

8. Since the jammer can eliminate two frequencies at a time, only half of the frequency cells can be used for the process gain:

$$10\log(10) = 10 \text{ dB}$$

9. The only difference is that the amplitude of the two phasors have two amplitude positions besides the phase shifts producing 4 times as many resultant phasors.

Answers

10. The only difference is that the π/4DQPSK only uses four phase states (±45, ±135) and D8PSK uses all eight phase states (±45, ±135, ±90, 0, 180) on the phasor diagram.

Chapter 3:

1. A circulator or a T/R switch.

2. $G_r = 20\log[V_o/2^n] - \text{MDSI} = 20\log[1/2^8] - [-114 - 10\log 10 + 3 + 4] = 68.8$ dB

3. Third Order SFDR = $2/3(\text{IIP}_3 + 174 - \text{NF} - 10\log B) = 2/3(20 + 174 - 3 - 10 \log(10 \text{ MHz})) = 80.66$ dB.

4. $80.66 - 10 = 70.66$ dB.

5. 3 dB, assuming no losses before the LNA.

6.   1X0 = 10 MHz
    0X1 = 12 MHz
    1X1 = 2 MHz, 22MHz
    1X2 = 14 MHz, 34 MHz
    2X1 = 8 MHz, 32 MHz
    3X0 = 30 MHz
    0X3 = 36 MHz

7. $120/100 = 1.2$.

8. According to the Nyquist criteria, Sample rate = $2(2\text{MHz}) = 4$ Msps.

9. Maximum phase error = $45 - \tan^{-1}(1/2) = 18.43$
   Maximum amplitude error = 6 dB

10. The advantages for oversampling the received signal

Answers

is better resolution for determining the signal and better accuracy for determining the time of arrival.

The disadvantages are increased sampling rate is generally more expensive if the sampling rate can be achieved and requires more processing for the increased number of samples taken.

Chapter 4:

1. Voltage controlled attenuator and a variable gain amplifier.

2. The RC time constant should be much larger than the period of the carrier and much smaller than the period of any desired modulating signal. The period of the carrier is .1 µs and the desired modulating signal period is 1 µs. Therefore, the period should be about .5 µs.

3. The integrator provides a zero steady state error.

4. As the error approaches steady state (very slow changing error), then the gain of the integrator approaches infinity. Therefore, any small change in error will be amplified by a very large gain and drives the error to zero.

5. a. The resistor does not affect steady state error since the gain of the op amp at low frequency or steady state is virtually unchanged. The gain is more for high frequencies or fast responses.

   b. The resistor reduces the gain in the steady state so that the error is no longer zero, and the less the gain (smaller the resistor) the more steady state error is present.

6. If the slope is non-linear, using a linear

Answers 405

approximation causes that in some operational periods the loop gain is less or more than the approximation and the response time will be slower or faster respectively.

7. The diode is non-linear and the point where the piecewise linear connections are, the diode provides a smoothing function due to the non-linearity.

8. DC offsets in the AGC does not matter in the sense that there is an infinite gain from the integrator. The AGC wants maximum gain with no signal input. The PLL however, with no signal input and a DC offset will tend to drift out of the lock range of the PLL.

9. They both are feedback systems. They just have different parameters that are in the feedback loop.

10. The PLL is in the lock state. This analysis does not include the capture state where the PLL has to search across a wider bandwidth to bring it into the lock state.

Chapter 5:

1. Weights need to be time reversed and are ± 1 not 0.
   x1 = 1, x2 = 1, x3 = –1, x4 = –1, x5 = 1, x6 = –1, x7 = 1.

2. $[Cos(\omega t + (0,\pi))]^2 = Cos^2(\omega t + (0,\pi)) = 1/2[1 + Cos(2\omega t + 2(0,\pi))] = 1/2[1 + Cos(2\omega t + (0,2\pi))] = 1/2[1 + Cos(2\omega t + 0)]$. Therefore the phase ambiguity is eliminated. However, the frequency needs to be divided in half to obtain the correct frequency.

3. BW = 2(50Mcps) = 100 MHz.

4. (a) The bandwidth is unchanged.
   (b) The bandwidth is reduced by the process gain.

5. Intersymbol interference is the distortion a chip which interferes with the other pulses, usually with the adjacent pulses.

6. Center of the eye. The transition point of the eye.

7. (a) Signal would need to be squared 3 times.
   (b) Signal would need to be squared 4 times.

8. (a) 45 degrees.
   (b) 22.5 degrees.

9. A two-frequency discriminator could detect the MSK waveform.

Answers

10. The MSK waveform sidelobe levels are significantly reduced.

Chapter 6:

1. Since the integral of the Probability Density Function is equal to $1 = 100\% = P_{wrong} + 37\%$; $P_{wrong} = 63\%$.

2. $E[x] = \sum x f_x(x) = 1(.4) + 2(.6) = 1.6$. The mean is equal to the $E[x] = 1.6$.

3. Closer to 2 because there is a higher probability that the answer is going to be 2 than 1.

4. $E[x^2] = \sum x^2 f_x(x) = 1(.4) + 4(.6) = 2.8$.

5. $\text{Var} = E[x^2] - \text{mean}^2 = 2.8 - 2.56 = .24$.

6. $\text{Std Dev} = \sqrt{\text{var}} = .49$.

7. $1 - .954 = .046 = 4.6\%$.

8. Double the clock frequency or sampler.

9. The probability receiving 1 pulse is .98. The probability of receiving all 20 pulses is:

    $.98^{20} = 66.8\%$.

10. Using the binomial distribution function the probability of only one error is:

Answers

$$p(19) = \binom{20}{19} p^{19}(1-p)^{(20-19)} = \binom{20}{19}(.98)^{19}(.02)^{1}$$

$$= \frac{20!}{(20-19)!19!}(.98)^{19}(.02)^{1} = 26.7\%$$

Therefore, the percentage of the errors that result in only one pulse lost out of 20 is:

Percent(one pulse lost) = 66.8%/(100%–26.7%)
= 48.9%.

11. Since the chipping rate is 50 Mchips/sec(Mcps), the minimum pulse width would be 1/50 Mcps. The null-to-null bandwidth = 2/PW = 2/(1/50 Mcps) = 2(50 Mcps) = 100 MHz wide.

Chapter 7:

1. Glint errors are angle of arrival errors and scintillation errors are amplitude fluctuations.

2. Specular multipath affects the solution the most since it is more coherent with the signal and directly changes the amplitude and phase. Diffuse multipath is a constantly changing more random signal which looks more like noise and is usually smaller in amplitude.

3. The effect of multipath at the pseudo-Brewster angle for vertically polarized signal is very small, basically negligible.

4. There is basically no effect of multipath at the pseudo-Brewster angle for horizontally polarized signal. Therefore, the multipath still highly affects the incoming signal.

5. Rayleigh criteria.

6. The Rayleigh criteria is as follows:

$$h_d \sin d < \frac{\lambda}{8}$$

where:

   hd = peak variation in the height of the

Answers

surface.

d = grazing angle.

If the Rayleigh criteria is met, then the multipath is a specular reflection on a rough surface. Therefore:

$$10\sin(10) = \frac{\lambda}{8}$$
$$\lambda = 80\sin(10) = 13.89 \text{ meters}$$
$$f = \frac{c}{\lambda} = \frac{3 \cdot 10^8}{13.89} = 21.6 MHz$$

7. The divergence factor is the spreading factor caused by the curvature of the earth. This factor spreads out the reflecting surface. This is generally assumed to be unity since the effects are negligible for most applications, except for possibly satellites.

8. Leading edge tracking. The multipath returns are delayed from the desired return, so the radar detects the leading edge of the pulse and disregards the rest of the returns.

9. The vector addition is shown in the figure:

10. The power summation uses power instead of voltage. The reflection coefficient is squared to represent a power reflection coefficient.

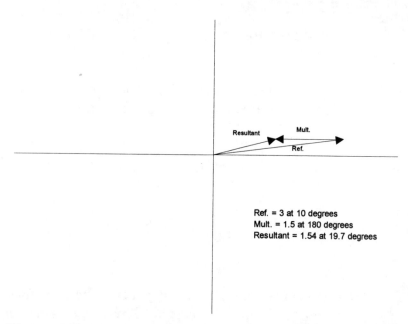

Figure P7.9 *Vector addition showing multipath effects.*

Chapter 8:

1. Jams at a rate equal to 1/response time of the AGC.

2. 1 MHz.

3. The adaptive filter uses feedback to update or adjust the weight values.

4. The unwanted sidebands during the mix down and up need to be eliminated.

Answers 413

5. The u value is the gain of the feedback process.

6. Increasing the u value does the following:
    1. Increases convergence time.
    2. Generally decreases stability.
    3. The steady state error is larger.

7. A digital system is discrete and the analog system is continuous. This becomes important in this adaptive design because with the digital system, everything is generally tied to a clock or synchronized. With the analog system, the waveforms are continuous and not tied to a clock. Therefore, the continuous delays need to be compensated for in the design.

8. The assumption is good when the jammer is much larger that the signal, and the directional antenna has significant gain towards the desired signal and provides a reduction in J/S.

9. The assumption is bad when the jammer is not larger that the signal and the directional antenna does not provide a reduction in J/S.

10. Channelized receiver. This receiver covers the largest instantaneous bandwidth with the best sensitivity.

Chapter 9:

1. C/A code = 2 × 1.023 MHz = 2.046 MHz
   P code = 2 × 10.23 MHz = 20.46 MHz.

2. Theoretical process gain for C/A code using the data rate of 50 Hz = 10 log 1.023MHz/50 = 43 dB.

   Theoretical process gain for P code using the data rate of 50 Hz = 10 log 10.23MHz/50 = 53 dB.

3. Short length code for faster acquisition times. Slower rate code for a higher signal to noise in a smaller bandwidth.

4. Long length code providing a more covert signal for detection. Faster rate code for higher process gain against unwanted signals.

5. The narrow correlator is better accuracy of the GPS solution.

6. a. The narrow correlator is less stable.

   b. The narrow correlator is easier to jam due to the wider bandwidth required to process the signal.

   c. The narrow correlator provides no benefit using a P code receiver due to the fact that the bandwidth is already limited in the transmitter.

Answers

7. a. Jitters the clock timing.

   b. Distorts ephemeris data regarding the orbits of the SV's.

8. Carrier smoothing is using the carrier change of phase with time to filter the code measured range data with time.

   Carrier data is not affected by the SA, multipath, and ionospheric effects as much as the code is affected.

9. Common errors, for example ionospheric errors, exist in both receivers and are subtracted out in the solution.

10. Wavelength ambiguity and cycle slips.

11. The wavelength of the difference frequency is larger, therefore there is less wavelength ambiguities to search over.

Chapter 10:

1. $\lambda = 3 \times 10^8 / 1 \times 10^9 = .3$ meters

   $\theta = \text{asin}[dp/((2\pi d)/\lambda)] = \text{asin}[10/((2\pi 3)/.3)] = 9.16$ degrees.

2. The elevation angle affects the azimuth interferometer calculation.

3. The direction cosines are defined for the cos α from the top down. Therefore, the angle up is the sin θ where θ is the elevation angle from the horizontal baseline.

4. The order is critical. There is a difference in the solution for the order of the coodinate conversion. For example, if the platform is pitched and then rolled, this gives a difference solution than if the platform is rolled and then pitched.